Solving S.H.I.T. Problems

Empowering Teams to Lead Through Crisis

Dr. James Ezhaya

Praise for *Solving S.H.I.T. Problems* by Dr. James Ezhaya:

"After decades of experience and millions of dollars invested in problem-solving trainings and certifications, the industry has not fundamentally improved the effectiveness of the root cause analysis process. Dr. Ezhaya's framework is revolutionary. It provides the missing piece, empowering professionals to achieve truly effective problem-solving."
    Agustin Lopez Diaz, Multinational Quality Executive and Global Industry Thought-Leader

"Dr. James Ezhaya's life intensity blazes, providing clarity and simplicity to take the right path, inducing courage and faith assuring the right road when addressing the most difficult problems. Results will follow, uplifting to the next level, transforming us and thus our organization. Are you ready for him?"
    Ruel Dato-on, Global Head of People Excellence, Siemens Energy

"With this book Dr. James Ezhaya takes practical problem-solving to the next level. It's a practitioner's guide for when to change problem solving strategy, providing clear guidelines and checklists. This book is a must read for people using traditional problem-solving techniques and who want to step up their game."
    Anders Helbo Mortensen, Head of Nacelle Development, Siemens Energy

"I've been in many S.H.I.T. scenarios with many companies (sometimes with Dr. James Ezhaya) and I've experienced the shortfalls of traditional problem-solving described in this book. I really enjoy how the T.R.U.S.T. model is laid out and focuses on the leadership aspects required for success. The leadership concepts described are truly the way leaders of the future must lead and the lessons are portable to all aspects of leadership."
    Landon Boyer, Vice-President Engineering, Schneider Electric

"James draws upon his experience across many industries along with detailed research to present, not only a framework to address an organization's most critical problems, but also practical tools including templates that can be applied."
        David Biederman, Senior Director, Operational Excellence at Carrier

"Dr. Jim Ezhaya's emerging and groundbreaking problem-solving technique goes beyond traditional approaches and allows to cope with severe situations rapidly and effectively. This methodology has been field-proven in automotive applications and enables dealing with unpredictable crisis by taking human factors into account. A must read for anyone who wants to mitigate and anticipate risk thru forward-looking team preparation."
        Benjamin Thoma, Vice-President, Faurecia

"When working with advanced and sophisticated technology and state-of-the-art manufacturing supply chains, the capability to problem-solve and learn from mistakes is key. Dr. Ezhaya's framework imbues skills to succeed in navigating professional challenges where others fail. These leadership skills will stand out among peers who continue to use outdated approaches. I highly recommend this book!"
        Wojciech Ostrowski, Chief Operations Officer, General Electric

"Learnings from the private into the public sector is a gift. This book is a rarity in its richness. Dr. Ezhaya deftly interprets lessons from failures shared by Fortune 500 leaders around the globe. From the ashes of past disasters, he draws a roadmap for the future that professionals at any stage of their career can use."
        Dr. Saeed Mubarak Kharbash Al Marri, Chief Executive Officer - Arts & Literature, Dubai Culture & Arts Authority

"A unique and fulfilling, must-read by Dr. J., who uncovers reasons behind ineffective problem-solving that have been ignored by others. He does a great job of explaining the soft reasons behind failed problem-solving practices in a clear-cut and structured way and proposes an effective way to tackle those areas to make problem-solving a success again."
        Dr. Yavuz Goktas, Industry Thought-Leader and Pioneer in Reliability and Design for Six Sigma

ACKNOWLEDGEMENTS

I am extremely fortunate and grateful to have worked as a professional in problem-solving at outstanding companies, which trained me to learn and use industry tools and methods: 8D leader and coach; Six Sigma Master Black Belt; Prince2 and PMP Project Manager; Stanford University Certified Project Manager, etc. For years I held the rules of these industry "truths" to be self-evident. Decades later, during my doctoral studies at the University of Warwick, I was empowered to challenge these commonly held truths of practice. My supervisor, Dr. Nicola Burgess, a leading academic in Lean for Healthcare, supported me tirelessly for 7 years. This journey would not have been possible without her support. Dr. Burgess is part of a world-class faculty and support team at the University of Warwick Business School, a long list of people who helped, challenged, urged, and encouraged me over my seven-year journey.

I am grateful to the 24 senior industry leaders from around the globe who shared their experience and insights. I cannot name you, but you know who you are. Your bravery and willingness to examine past failures, will benefit future generations. Thank you for your advice and support.

I am humbled, by the outpouring of support I received from friends and colleagues, especially the multiple reviewers of my manuscript who made time to read, challenge, and improve this body of work.

Ideas aren't action. If the ideas in this book are good ones, they will need to be put into practice and tested. And when that happens, we'll learn more. This is how we will all improve. If you want to be part of that, please connect with me online at https://theshitshow.biz. Thank you in advance for your feedback, suggestions for improvement, your stories, and insights on managing S.H.I.T. and using the T.R.U.S.T. framework. Let's grow the community of professional problem-solvers together!

This book is dedicated to my children.

Thank you for your love and inspiration. I dedicate this book to you because I love you.

And also to make up for the fact that for the rest of your life, your name will be synonymous with "S.H.I.T."

Sorry,

Dad

Contents

INTRODUCTION A problem we all face 1

    Book Structure 3

    The Goals 4

CHAPTER 1 Why are we here? A brief history of problem-solving 6

    The Second World War accelerated adoption of formal problem-solving methods 7

    The Third wave of statistical control: the 1980s 9

    Current problem-solving requirements in engineering industries 11

PART 1 15

CHAPTER 2 Why Traditional Problem-Solving Fails 16

    Academic critique of traditional problem-solving 17

        Criticism 1: Traditional problem-solving is 'one-size-fits-all' 17

        Criticism 2: Traditional problem-solving leaves knowledge on the table 19

        Criticism 3: Traditional problem-solving fails to account for human dynamics 21

    Practitioner Critique of Traditional Problem-Solving 22

        Concept 1: Social Response to Pressure 24

        Concept 2: Social Response to People 29

        Concept 3: Social Response to Politics 36

    Chapter conclusion and next steps 42

CHAPTER 3   When Traditional Problem-Solving Fails: S.H.I.T. 43

   Case 1: Product Design Failure led to Power Blackouts in Africa 50

   Case 2: Steam Turbine Scrapped Component 57

   Case 3: Wind Turbine Service Technician Fatality 65

Chapter conclusion and next steps 74

PART 2 75

CHAPTER 4   The Solution: T.R.U.S.T 76

Understanding human response to severe events through the lens of Sensemaking 77

   Leaders 81

   Communication 90

   Cross-functional experts 94

   Enactment: creating knowledge through action 99

Improving human response with Psychological Safety 105

   Inclusive Leadership: enabling voice to minimise the impact of power differences 108

   Interdisciplinary action teams 112

   Successful teaming: promote equality and stability to draw out technical expertise 114

   Team touchpoints: communication needed for real time learning 116

   Tacit knowledge: working on the edge of codified expertise 116

Synthesis of lessons from Sensemaking and Psychological Safety .. 118

Idea to action 1: Leadership 119

Idea to action 2: Design of communication structures 120

Idea to action 3: Empowered cross-functional technical experts . 121

Idea to action 4: Enactment – learning by doing 122

Idea to action 5: Framing for learning - Social Construction of Reality 123

CHAPTER 5   T.R.U.S.T. for Practitioners, Step-by-Step 125

Tool 1. S.H.I.T. Problem-Solving Identification Evaluation 128

   Roles and Responsibilities 129

   Process 129

   Tool Overview 131

Tool 2. Problem-Solving Model to Build T.R.U.S.T. 134

   Roles and Responsibilities 134

   Process 136

   Tool Overview 138

Tool 3. Situational Analysis – Frame the Problem – 5W2H 142

   Roles and Responsibilities 143

   Process 143

   Tool Overview 143

Tool 4. Severe Event Alert 147

   Roles and Responsibilities 147

   Process 148

Tool Overview 148

Tool 5. Permission Charter to Promote Psychological Safety 152

    Roles and Responsibilities 152

    Process 152

    Tool Overview 153

Tool 6. T.R.U.S.T In Action Checklist 154

    Roles and Responsibilities 154

    Process 154

    Tool Overview 155

Chapter conclusion and next steps 157

CHAPTER 6   T.R.U.S.T. in Practice 158

Genesis of a S.H.I.T. Problem - RED ALERT ISSUED 160

    8D problem-solving tried unsuccessfully 163

    Additional external pressure and complexity: COVID-19 166

    S.H.I.T. problem conditions confirmed using the T.R.U.S.T. solutions 167

T.R.U.S.T. principles applied 169

    (T) Tsar appointed independently to govern and support the problem-solving team 171

    (T) A team of empowered technical experts is established 173

    (S) Support open, streamlined communication structures 178

    (R) Re-frame failure as an opportunity for knowledge 183

(U) Understand and define the scope of the problem 186

CHAPTER 7   CONCLUSION 194

BIBLIOGRAPHY 197

APPENDIX   Research Methods 220

   Design Science Research: the combination of practice and theory .. 220

   DSR and CIMO-Logic for field research 221

   Interview approach and study participants 222

      Semi-structured interviews 222

      Interview research method application 224

      Data gathering – Semi-structured interview protocol and process 224

      Limitations of the interview method 226

      Study participants and severe event identified 227

   Data analysis 235

Table of Figures

Figure 1 Military Standard 1520 - normalising technical problem-solving 9
Figure 2 The logic filters 10
Figure 3 Problem-solving methodologies are codified in international standards 11
Figure 4 AIAG CQI-21 Problem Solving and Complexity 13
Figure 5 Aggregation of first and second order concepts and themes: Pressure 28
Figure 6 Aggregation of first and second order concepts and themes: People 35
Figure 7 Aggregation of first and second order concepts and themes: Politics 41
Figure 8 Concepts, themes, and factors leading to Severe 45
Figure 9 Concepts, themes, and factors leading to Highly Visible 46
Figure 10 Concepts, themes, and factors leading to Information deficient 47
Figure 11 Concepts, themes, and factors leading to Time-Sensitive 48
Figure 12 Example of wind turbine failure mode - fire 66
Figure 13 Sensemaking during a severe event 81
Figure 14 STICC method for sensemaking in a crisis situation 88
Figure 15 How T.R.U.S.T. Works 125
Figure 16 S.H.I.T. Problem Identification Process 130
Figure 17 S.H.I.T. Problem-Solving Identification Evaluation 132
Figure 18 Process flow group model for solving S.H.I.T. problems 137
Figure 19 Problem-Solving Model to Build T.R.U.S.T. 139
Figure 20 Frame the problem using 5W2H, step-by-step 144
Figure 21 Template for Situational Analysis Using 5W2H 147
Figure 22 Process for creating the severe event alert 149
Figure 23 Severe Event Alert Template 150
Figure 24 Permission Charter to Promote Psychological Safety 153
Figure 25 T.R.U.S.T. In Action Checklist 156
Figure 26 Autonomous Vehicle Display Panel 159
Figure 27 Timeline Day 1: Major events during the S.H.I.T. problem 163
Figure 28 S.H.I.T. Problem-Solving Identification Evaluation Tool completed 168
Figure 29 Model for solving S.H.I.T. problem using T.R.U.S.T. principles 170
Figure 30 Checklist developed for S.H.I.T. problems 191
Figure 31 DSR process flow 221
Figure 32 Interview overviews 234

INTRODUCTION A problem we all face

I am a quality guy, an operations professional in the engineering and manufacturing sphere. When a product or service fails to perform as planned, we are the people called in. Problems related to product design, safety, production, quality, and logistics failures, aka 'technical problems', usually fall to the quality professionals to solve. To solve these problems, practitioners world-wide are required to apply root cause analysis problem-solving techniques with very official sounding names such as Six Sigma Breakthrough Strategy or the Eight Disciplines of Quality (8D). This approach to problem-solving is intended to determine the root cause of a problem in order to identify and implement improvement actions through direct and sequential analysis, deductive logic, and advanced statistical methods.

In this role, I have seen death, permanent disability, fire, massive product failure, and more. These types of problems appear all around us and well-known examples are easy to recall: Boeing 737 Max planes crashing;[1] Samsung batteries catching on fire;[2] Tesla vehicles fatally driving head-on into traffic;[3] an Orlando amusement park ride[4] fatally ejecting a 14-year old boy; a condo building in South Florida collapsing and killing all

---

[1] https://www.nytimes.com/2020/11/24/sunday-review/boeing-737-max.html
[2] https://www.nytimes.com/2017/01/22/business/samsung-galaxy-note-7-battery-fires-report.html
[3] https://www.nytimes.com/2021/07/05/business/tesla-autopilot-lawsuits-safety.html
[4] https://www.nytimes.com/2022/04/26/us/florida-amusement-park-ride-death-lawsuit.html

occupants[5]; Johnson & Johnson destroying 60 million COVID vaccine vials in the middle of the pandemic.[6]

Trained in the most advanced statistical, logic-based problem-solving methods in practice, it has been my job to lead teams through these types of disasters. Despite decades of professional experience and training, I found that the industry standard tools and approaches often did not work. In fact, sometimes these standard methods were manipulated to derail the investigation and hide information rather than form the basis of a solution. And none of my training helped address the emotional chaos experienced by personnel touched by these types of tragedies.

I was not alone.

For seven years, as part of a doctoral research project, I interviewed dozens of senior level leaders from large, multinational companies to learn from their experience on problem-solving. All were trained and had extensive experience using industry standard approaches to technical problem-solving. All had faced serious, real-world problems and had seen the standard methods fail. These failures had led to significant losses, financial costs, and emotional tolls on their teams. Looking across all these lived narratives, I saw patterns emerge. There were different leaders in different companies and in different countries, but all with the same problem: industry standard approaches to problem-solving had failed and no one had an answer.

---

[5] https://www.nytimes.com/2022/06/23/us/surfside-condo-collapse-alarm.html
[6] https://www.nytimes.com/2021/06/11/us/politics/covid-vaccine-emergent-johnson-johnson.html

This book identifies these repeating patterns of failure affecting companies worldwide and applies cutting edge academic research with real-world practitioner experience to develop a new approach to problem-solving for severe events. Readers will learn how to identify severe technical problems where traditional problem-solving methods will fail and thus will be equipped to lead and succeed when facing severe problems. The book begins with an overview of current methods and their origins, then coaches readers to acquire two distinct and valuable new skills:

1. Predict and identify problem situations where traditional problem-solving methods are likely to fail.

2. Develop and deploy a response to these types of problems that will reach successful resolution.

The appendix outlines and details the academic research methodology applied.

Book Structure

Part 1 explains the conditions that, when present, frequently cause traditional problem-solving to fail. These conditions come from analysis of 27 unique, actual cases along with contributions from industry professionals. The examples revealed four key problem characteristics that cause traditional problem-solving methods to break down: the problem is Severe, Highly visible, Information-deficient, and Time sensitive. When all these conditions are present, you have a S.H.I.T. problem, and the typical approach to problem-solving is likely to fail. You need something new.

Part 2 introduces a new approach to augment traditional problem-solving in order to mitigate the impact of S.H.I.T. conditions. Drawing on the knowledge of industry experts from around the world, leading academic scholarship, and original research, I have devised a practical solution that professionals can use today to turn S.H.I.T. problems into wins: the T.R.U.S.T. approach to problem-solving. Part 2 will explain this new method and equip readers with the tools to use it successfully.

The Goals

More than 100 years have passed since the linear, statistical approaches to problem-solving commonly in use today were developed, and the world has changed in that time. Today's problems occur in a significantly more complex and multi-layered environment of liability, litigation, and 24-hour news cycles. In addition to these pressures, manufacturers are working with rapidly evolving technologies to accelerate the innovation process for designing new products, all while competing globally to survive. Problems that arise in this current environment require an approach that can adapt established problem-solving methods to meet contemporary needs. This book will equip operations and quality personnel with the knowledge and tools to augment their traditional approaches to problem-solving and succeed where old methods have failed. My research aims to draw wisdom from tragic events faced by those that came before in order to empower leaders so that they may minimize the suffering and other harmful impacts these tragedies engender.

> 'He who learns must suffer. And even in our sleep pain that cannot forget falls drop by drop upon the heart, and in our own despair,

> against our will, comes wisdom to us by the awful grace of God'.
> [Aeschylus 458 BC][7]

In misfortune, writes the Greek poet Aeschylus, comes wisdom. Professionals that recognise this are able to achieve first-order resolution of the problem they face, and also capture and operationalise second-order learning to avoid future tragedy. My goal is to empower the reader with knowledge to respond to tragedy in a way that will minimise negative impacts and accelerate learning.

---

[7] Aeschylus (525-456 BC) Greek dramatist (Æschylus) Agamemnon, ll. 175-183

CHAPTER 1 Why are we here? A brief history of problem-solving

Technical problem-solving is often referred colloquially as 'root cause analysis', and is a linear problem-solving process, employed in order to understand, identify, and implement improvement actions. The American Quality governing body defines root cause analysis as: 'the method of identifying the cause of a problem, solving it, and preventing it from occurring again. Uncovering the correct and accurate reason(s) why something is happening or has already occurred'.[8]

Technical problem-solving originates from research and experiments performed over 100 years ago by the manufacturing arm of the Bell System AT&T in the electrical-engineering business of Western Electric. At the celebrated Hawthorn Works plant, statistician Dr. Walter Shewart developed the engineering problem-solving method known as Statistical Process Control (SPC)[9] in the 1920s. Adapting the principle developed by American contemporary Frederick Winslow Taylor, 'science, not rule of thumb',[10] SPC quantitatively measures the distribution patterns in a process in order to identify and separate common cause variations from special cause variations. SPC defines when a variable exceeds the statistical limits of common cause variation, usually set at standard deviations from the production average (e.g., +/- 3-sigma), which signals the presence of non-random, special-cause, exceptional variation and

---

[8] Source: ASQ Quality Dictionary https://asq.org/quality-resources/quality-glossary
[9] Best and Neuhauser 2006
[10] Locke 1982

triggers the application of root cause analysis to identify the assignable source(s) and apply remedial actions.[11] This method of problem-solving evolved the process itself such that it became a source of learning, variation reduction, and continuous improvement.

It is noteworthy that the beginnings of SPC and problem-solving have been credited to the three alumni and colleagues from Hawthorn works (the Bell Technologies plant): Dr. Walter Shewart,[12] Dr. W. Edward Deming, and Dr. Joe Juran. These forefathers of quality and industry are often referred to by honorary titles: Dr. Shewart as the 'father of Quality Control', Dr. Deming as the 'father of Quality Improvement', and Dr. Juran the 'father of Quality Assurance'.[13] Medals are awarded annually in their respective names for outstanding contributions to the field of Quality, Statistical, or Executive leadership contributions.[14]

The Second World War accelerated adoption of formal problem-solving methods

After the bombing of Pearl Harbor brought the United States formally into World War II, a War Production Board was created. The War Production Board coordinated the domestic supply chain to enlist industries such as automotive and aerospace into the American wartime efforts. The U.S. government secured personnel from the Bell system and Bell system industrial methodologies were codified and secured as military (MIL)

---

[11] Deming 1975a; Deming 1975b
[12] https://asq.org/about-asq/honorary-members/shewhart, Dr. Walter Shewart, the father of statistical process control
[13] https://asq.org/about-asq/honorary-members
[14] https://asq.org/about-asq/asq-awards/shewhart

standards, which are still in use. SPC methods became a mandatory element in military contracts and the global supply chain.

In 1974 the U.S. military released MIL-STD-1520, 'Corrective Action and Disposition System for Nonconforming Material', expanding the quality standards to include establishment of the concept of containment actions for nonconforming items and corrective actions for lessons learned and improvement. In addition, the MIL-STD-1520 provided a common lexicon for understanding concepts related to technical problem-solving and nonconformity management.

Figure 1 presents the cover for Military Standard 1520. This document states that it is insufficient for operations and quality personnel to classify the nonconforming material (e.g., scrap or rework) without applying corrective action problem-solving, learning, and improvement. These elements are now part of the ISO 9000 lexicon and of everyday practice globally.

MILITARY STANDARD
Corrective Action and Disposition System for Nonconforming Material

NO DELIVERABLE DATA
REQUIRED BY THIS DOCUMENT

AREA QCIC

Figure 1 Military Standard 1520 - normalising technical problem-solving

The Third wave of statistical control: the 1980s

The early 1980s continued the maturity and growth of technical problem-solving into what is referred to as the Third Wave of Statistical Quality Control,[15] when today's mandatory methods such as Six Sigma and 8D were introduced, popularised, and rapidly applied. First introduced in 1982, the innovation of Six Sigma incorporated tools of SPC organised as a project management method to create a scientific data-driven system of problem-solving. Originally named the 'Logic Filters' process, Six Sigma is visualised as a funnel where the application of the statistical toolset at each phase of the process is used to identify the root cause(s) of a problem.

---

[15] Juran 1995

Figure 2 shows the problem-solving phases of the original 'logic filters' approach.[16]

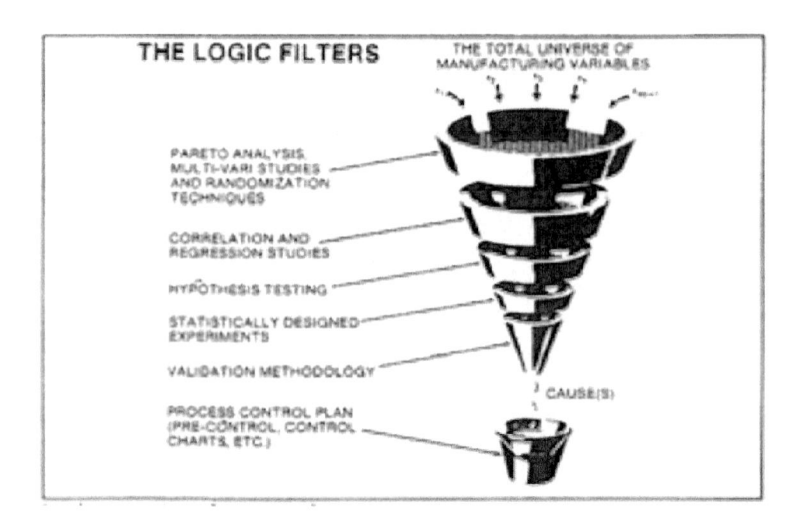

![The Logic Filters diagram showing an inverted funnel with phases: Pareto Analysis, Multi-Vari Studies and Randomization Techniques; Correlation and Regression Studies; Hypothesis Testing; Statistically Designed Experiments; Validation Methodology; Process Control Plan (Pre-Control, Control Charts, etc.). The total universe of manufacturing variables flows in at the top, narrowing down to Cause(s).]

Figure 2 The logic filters[17]

'SPC Experts' were qualified and certified as practitioners in the process through an apprenticeship program. By the mid-1980's 'Logic Filters Problem-Solving' strategy was renamed 'Six Sigma breakthrough problem-solving' strategy. The 'SPC Experts' nomenclature was adapted to several levels of SPC competence renamed to parallel the belt levels in karate (e.g., yellow, green, and black belts; master black belts). During this same time period, W. Edwards Deming created and introduced the

---

[16] https://www.mikeljharry.com/milestones.php, Archives for Mikel Harry, one of the founders of Six Sigma
[17] Source: Dr. Mikel Harry Archives (1985) Logic Filters Focus Curriculum https://www.mikeljharry.com/milestones.php , accessed October 16, 2021

approach known today as the Eight Disciplines of Quality, or 8D, at Ford Motor Company.

Over the next 30 years technical problem-solving matured and has been codified into a body of knowledge with processes, qualifications, and tools standardised into national (ANSI) and international (ISO) regulatory and legal documents. Adherence to these international standards are the proof and reference documents for the ISO certification requirements in applying corrective and preventive actions to nonconformities. Figure 3 shows a sample of the various technical problem-solving methods which have been released as international standards.

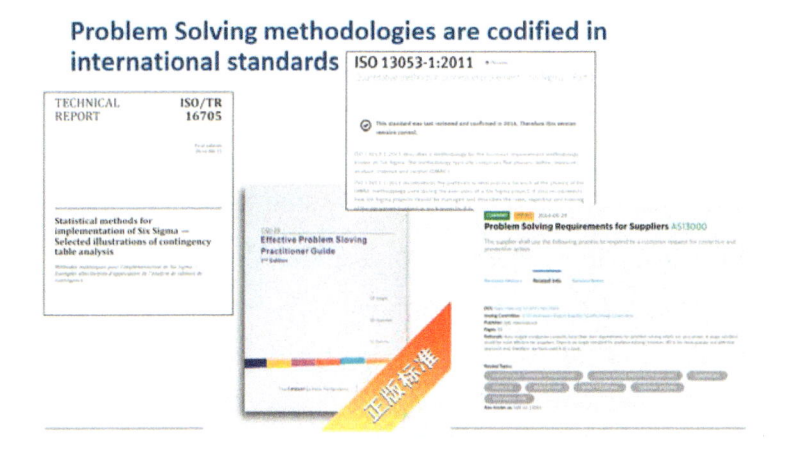

Figure 3 Problem-solving methodologies are codified in international standards

Current problem-solving requirements in engineering industries

To manage nonconformities, technical industries as diverse as manufacturers of wind turbines, jet engines, powerplants, and automobiles all require their suppliers to follow the methods of statistical process control. For industry practitioners, application of the formal problem-solving standards is not one of theory or choice: contractually, companies and suppliers are bound to follow the selected technical standard specified for problem-solving when managing a nonconformance.

It should be noted that at the company level, a firm may undertake and apply multiple ISO and industry standards based on the problem type and the complexity of technical non-conformance. For example, a recommendation from the Automotive Industry Action Group (AIAG) in its standard CQI-21 is shown in Figure 4: Effective Problem-Solving Leader Guide (version 1 issued 9/2012). For simple problems, the company can apply everyday Kaizen continuous improvement, but as the problem complexity increases and a team is needed, the company draws upon 8D or Six Sigma, using technical standards guiding their usage.

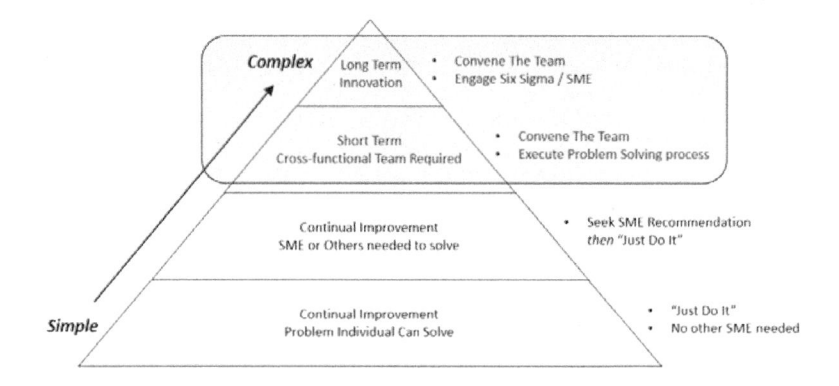

Figure 4 AIAG CQI-21 Problem Solving and Complexity

Despite 100 years of technical problem-solving being prescribed and enforced by manufacturers around the world, there are known limitations with the application of such methods. A report by Deloitte, in collaboration with the AIAG,[18] surveyed 22 automotive companies and their suppliers on the top issues affecting the automotive industry. The results were written and published in their report called 'Quality 2020: Automotive Industry's View on the Current State of Quality and a Strategic Path Forward'.[19] In this report, 'Concerns related to problem-solving' were ranked as the

---

[18] Founded in 1982, Automotive Industry Action Group (AIAG) is the automotive trade association. Its impact and membership extend beyond automotive to include over 5,000 member companies from consulting, government, academia, healthcare, and many other industries. Members include some of the largest firms in the world. A sample of the membership is available at https://www.aiag.org/supplier-connect.

[19] Deloitte available at https://www2.deloitte.com/us/en/pages/manufacturing/articles/automotive-quality-2020-report-aiag.html

most critical issue impacting automotive, that is, ranked higher than new product development, local customer requirements, and loss of experience.[20] The report stated: 'About 95 percent of respondents believe closing the gap in problem solving would have a moderate to extremely high impact on quality'.[21] The purpose of this book is to empower operations professionals with the knowledge and skills to close that gap.

---

[20] https://go.aiag.org/quality-2020-report, last accessed 27/1/22
[21] AIAG Quality 2020: 5.

PART 1

Goal: Predict and identify problem situations where traditional problem-solving methods are likely to fail

CHAPTER 2 Why Traditional Problem-Solving Fails

In order to be able to predict and identify when problem-solving will fail, I first sought to understand why it fails. I conducted an extensive literature review of hundreds of academic articles on technical problem-solving. I studied and parsed literature published over the last 30 years in leading academic journals such as *Academy of Management Journal*, *Management Science*, and the *Journal of Operations Management.* This exercise provided a robust background and theoretical context that helped me develop a lens to inform my assessment of real experience.

I then moved beyond theory to research of real-world crises experienced by practitioners in problem-solving. Over the course of seven years, I conducted interviews with senior leaders willing to share their problem-solving failures first hand. Their candour and bravery in sharing these experiences made it possible to examine 27 actual tragedies where traditional problem-solving was employed but didn't work. To identify common patterns, I employed rigorous qualitative data analysis to hundreds of pages of structured interviews transcribed and coded to distil emergent themes. This process revealed that certain predictable failure modes occur and recur when problem-solving fails. Across industries, geographies, companies, and countries, problem-solving fails during extreme conditions generated during a severe event.

Academic critique of traditional problem-solving

I'm not the first to conclude that today's approach to problem-solving has problems of its own. Scholarly critique identified ideas I had personally observed and heard repeated from other professionals and practitioners. Three themes stand out as particularly relevant:

1. Traditional problem-solving is 'one-size-fits-all'
2. Traditional problem-solving leaves knowledge on the table
3. Traditional problem-solving fails to account for human dynamics

Criticism 1: Traditional problem-solving is 'one-size-fits-all'

Evidence shows that as problem complexity and severity increase, social complexity also increases. Yet, there is a built-in assumption in typical problem-solving approaches that no matter how complex or severe a problem is, the method for solving the problem is a one-size-fits-all approach. Typical approaches to problem-solving don't include a response for navigating a more politically and socially complex environment. In other words, a technical problem-solving expert correctly applying the typical approach may enforce a narrow focus on the problem as a statistical problem, even when these statistical tools are not effective to deal with the social forces in play.[22]

---

[22] de Mast and Lokkerbol 2012

Scholars argue that technical problem-solving methods do not suffice for addressing complex problems, because problem-solving leaders are not trained to manage the political and social elements in a complex problem. And in the case of a severe event, the typical problem-solving process can become politically 'hijacked':

> [Technical problem-solving] reports themselves, influenced by the need to preserve interpersonal relationships and by hierarchical tensions and partisan interests, may not always reflect the content of discussions during investigations nor the realities of what happened.[23]

---

[23] Peerally et al. 2017: 418

Finally, in a study to investigate technical problem-solving effectiveness for complex problems, University of Pennsylvania Wharton Business School professor John Paul MacDuffie performed an analysis of problem-solving at three automotive companies (Ford, GM, and Honda). The study found that in the context of technical problems, 8D was treated as a tool for satisfying management:

> In general, the 8Ds appear to be used more to report on the activity level of the subsystem group, to show that the required processes are being fulfilled, rather than to diagnose, systematically, the 'root cause' and possible solutions to a problem.[24]

In the face of complex problems, the typical approaches sometimes become a 'check the box' exercise of writing 8D reports to satisfy company policy to meet internal company requirements.

Criticism 2: Traditional problem-solving leaves knowledge on the table

Technical problem-solving was born out of mass production in a manufacturing environment. When variation is detected beyond an acceptable, statistical sampling criteria, trained experts launch the 8D or Six Sigma project, using the data to mathematically attack the problem and gain new control over the process. However, researchers have identified that the typical methods ignore the need for other types of experts on problem-solving teams: 'An important limitation of the method is its

---

[24] MacDuffie 1997: 489

generality, which limits the methodological support it provides, and which fails to exploit task domain specific knowledge'.[25] Task domain knowledge, such as engineering experts, can bring crucial information learned through direct experience with the product and process. In typical problem-solving approaches, these folks aren't invited to be on the team, and all that knowledge is left on the table.

---

[25] de Mast and Lokkerbol 2012: 604

Criticism 3: Traditional problem-solving fails to account for human dynamics

Technical problem-solving has evolved over nearly a century to become highly standardised across the industrial engineering industry, but the dominant logic-based problem-solving methodologies fail to account for the human elements that can influence and even drive the problem-solving effort: 'In general, research on problem solving has focused on identifying preferable methods rather than on what happens when human beings confront problems in organisational contexts'.[26]

Processes designed with the 20th century scientific mindset were developed in the context of professional hierarchy and clear roles and responsibilities, in contrast to 21st century problems that occur within networked organisations, built on fluid and agile teams. Current methods are not fully sufficient and 'efforts to encourage second-order problem solving will require addressing psychological, organisational, and institutional factors, rather than any one of these in isolation'.[27] In severe event problems a situation can deteriorate rapidly and statistical based technical problem-solving is not designed for the soft power needs such as voice, team dynamics, and need for communication.

In sum, academic studies have noted the limitations of traditional and/or dominant logic-based problem-solving methods when employed to solve

---

[26] Tucker et al. 2002: 125
[27] Tucker et al. 2002: 134

severe technical problems, in part because they do not address the role of human behaviours during extreme conditions.

Practitioner Critique of Traditional Problem-Solving

In addition to a thorough review of academic literature on this topic, I reached out to expert practitioners to get their perspective on traditional problem-solving approaches. Though the group of contributing experts have a diverse, varied, and unique set of perspectives and experiences, they all share the following characteristics:

- Senior and/or executive leader at Fortune Global 500 firm
- Employed in the operations and/or engineering functions
- Involved personally in technical problems and problem-solving

Interestingly, the practitioners were not selected based on their experience with a severe event problem where traditional problem-solving failed. And yet, every single person had such an experience, often referred to as a 'career defining moment'. Several of them told me about more than one such instance.

The industry leaders that contributed to the solutions in this book represent senior leaders in a variety of manufacturing and production functions. Participants are involved with large global corporations that design, produce, and service manufactured goods. Contributors are from around the globe, working in the U.S., U.K., France, Poland, Germany, Switzerland, Denmark, Mexico, Japan, People's Republic of China, and Taiwan.

The real-world feedback from structured interviews was generated using a formal protocol, supported and vetted by academic experts and thought leaders. Feedback was transcribed into a template for analysis and comparison. This data was analysed following the 'case research in operations' method of turning qualitative data examples into cases to answer 'how and why' questions[28] prior to data coding and thematic analysis.[29] Analysis was iterative, involving refinements and revisions to the findings as they emerged over time. All transcripts were coded in the qualitative software tool NVivo[30] using a multi-step process. An initial open coding approach was followed by a more detailed axial coding method,[31] applying the 'Gioia method', and concluded with 'first-order' and 'second-order' evaluations aggregated into causal dimensions.[32]

The Gioia approach to qualitative analysis is an inductive approach consisting of the development of first-order codes, followed by the creation of second-order codes that lead to the aggregate dimensions. Using this approach, the expert feedback was sorted into three emergent concepts related to why problem-solving efforts failed: pressure, people, and politics, together referred to here as '3P'. These recurring 3P social responses undermine the problem-solving process.

---

[28] Voss et al. 2002
[29] Locke et al. 2022; Gioia et al. 2013
[30] NVivo is computer software used in qualitative data analysis to organise, analyse, and find insights in unstructured or qualitative data such as interviews. It is frequently used by academic researchers to perform deep levels of analysis on large volumes of data.
[31] Strauss and Corbin 1998; Scott and Medaugh 2017
[32] Gioia et al. 2013

Concept 1: Social Response to Pressure

It's well recognised that stress can trigger a variety of physical and mental harms, affecting productivity and performance. Interviews revealed that the problem-solving environment is generating pressure and its related harms, including sleep problems, physical pain, and lack of concentration, can often follow the worker home. The consequences of pressure can worsen as the severe event unfolds, leading to a breakdown of the systems the problem-solving teams need to successfully operate (e.g., interpersonal communication, trust, creative thinking, etc.). Employee fears of doing something wrong, of being blamed, of being found inadequate, can hijack the problem-solving process. As research has shown, human responses to high stress situations are biologically engrained. The concept of 'fight, flight, or freeze' is common vernacular to refer to the human default responses to high stress situations. Even experienced employees may become paralysed in their thinking, acting instinctively instead of thinking creatively or drawing on their knowledge, experience, technical expertise, and common sense.

The interviewees were at the senior or executive level of leadership at large multi-national companies, (i.e., accustomed to dealing with high-stakes challenges). Even so, when queried about problem-solving a major technical crisis, common themes of isolation, fear of failure, and being sad and depressed arose. Statements like 'we're all going to die' and 'it felt like the world was going to end' were conveyed. These responses were attributed to real and perceived threats of consequences and a strong sense of urgency to identify the root cause and solve the problem. As described by a Global Product Line Director:

> And, he says to me, 'you know, [NAME REMOVED], if this doesn't work, I'm gonna get the sack'. And, he goes, 'if it doesn't work, you're gonna get the sack as well - I'm gonna make sure you get the sack'. And, he went, 'when they're putting the last nail in my coffin, I'll think you're wrong for doing that'.

> Now, at that moment in time, you've got a [PRODUCT] that was late in manufacture, you've damaged it when you've picked it up… And, now he's saying, 'when you put the last nail in my coffin'. That's me finished here. Extra pressure!

An Engineering Director related how 'fix it or else' threats are used to increase pressure:

> And to reinforce actually, what it's like to have ownership of the problem-solving actions, I can describe a review I had with my boss's boss after the 8D review. While we actually walked out of this room with the business leader, I remember him actually turned to me, twice. First time he turned to me saying 'fix it.' And the second time he turned to me saying: 'Quick, or else!' It's hard to even describe in words, how pressured I felt at that particular moment. I have a family to feed and take care of.

And the threats proved to be very real in some cases. The interviews provided examples of how lack of progress with an investigation can lead to significant career impact, including disciplinary action and termination. A former CEO communicated the clear dichotomy between everyday problems and severe events in relation to pressure:

> As long as the stakes are rising, that eventually will sort of tear apart your confidence. …. if it is a small problem…. it can be solved and replaced fairly easily and at no cost or whatever. But as soon as you're talking about huge stakes, people behave very differently because of the pressure. It becomes you or me…

The interviews clearly identified a pattern of social responses to pressure during a traditional problem-solving effort. Figure 5 illustrates how the raw data from these interviews evolved into second order themes and terms which, through qualitative analysis led to the aggregate dimension that is the social response to pressure when problem-solving during a crisis.

Figure 5 Aggregation of first and second order concepts and themes: Pressure

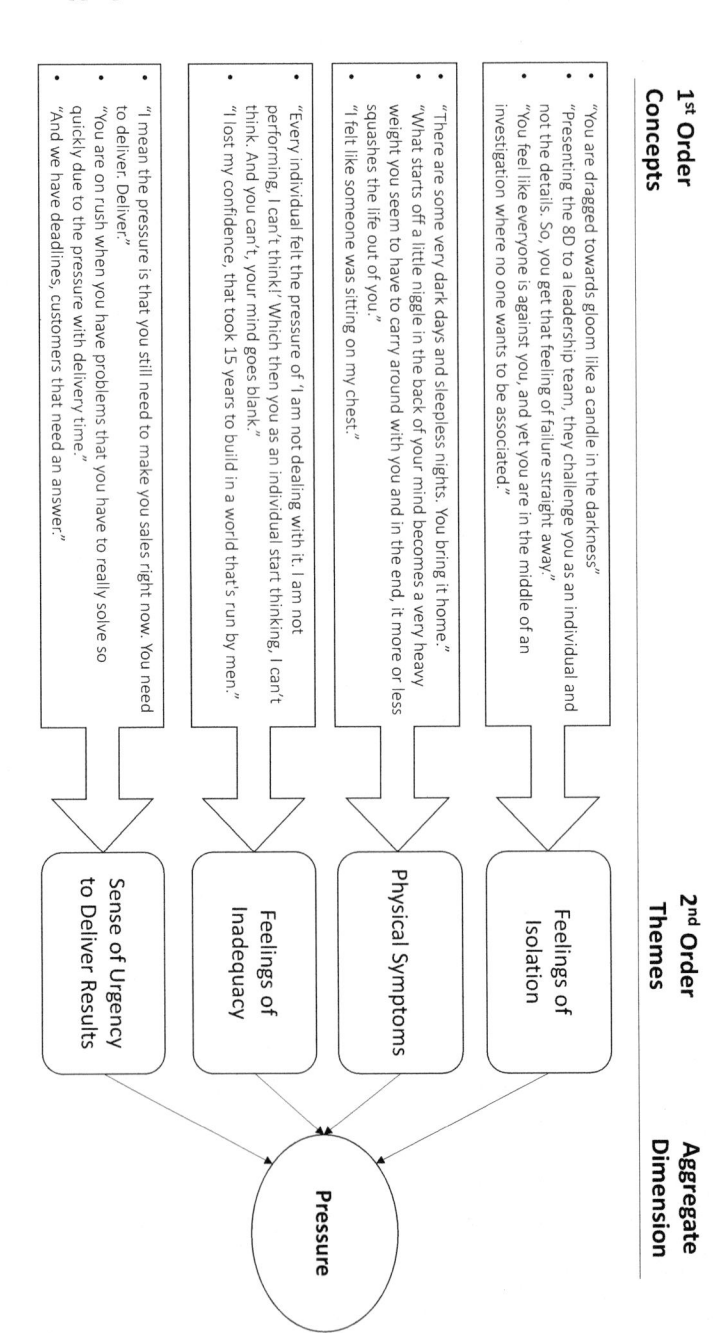

1st Order Concepts

- "You are dragged towards gloom like a candle in the darkness"
- "Presenting the 8D to a leadership team, they challenge you as an individual and not the details. So, you get that feeling of failure straight away."
- "You feel like everyone is against you, and yet you are in the middle of an investigation where no one wants to be associated."

- "There are some very dark days and sleepless nights. You bring it home."
- "What starts off a little niggle in the back of your mind becomes a very heavy weight you seem to have to carry around with you and in the end, it more or less squashes the life out of you."
- "I felt like someone was sitting on my chest."

- "Every individual felt the pressure of 'I am not dealing with it. I am not performing, I can't think!' Which then you as an individual start thinking, I can't think. And you can't, your mind goes blank."
- "I lost my confidence, that took 15 years to build in a world that's run by men."

- "I mean the pressure is that you still need to make you sales right now. You need to deliver. Deliver."
- "You are on rush when you have problems that you have to really solve so quickly due to the pressure with delivery time."
- "And we have deadlines, customers that need an answer."

2nd Order Themes

- Feelings of Isolation
- Physical Symptoms
- Feelings of Inadequacy
- Sense of Urgency to Deliver Results

Aggregate Dimension

- Pressure

Discussion: the effect of pressure on the problem-solving capability of team members

Data analysis reveals that colleagues outside the problem-solving team apply pressure in the context of a severe event, which promotes a stress response in individuals tasked with problem-solving: specifically, anxiety and depression responses that deplete confidence and promote brain fog, fear, and an inability to act or make a decision. The consequences of such responses were multi-fold, undermining rationality and significantly hindering the logical and linear thinking needed for effective problem-solving. Evidence also showed that the behavioural effects of high levels of pressure worsened over time as the problem event continued.

As the visibility of the problem grew, pressure increased, further reinforcing stress responses and contributing to reductions in mental capacity when critical thinking and complex problem-solving were most needed. Further, it was reported that many people responded to these conditions in unproductive or damaging ways, such as substance abuse, resignation from the company, and lasting physical and psychological suffering. The pressure from the problem impacts individuals, and those negative effects undermine and disrupt the problem-solving effort.

Concept 2: Social Response to People (interpersonal relationships)

Severe problems create strain on relationships between co-workers and management, heightening fears of interpersonal risk. People respond to these fears in a variety of ways that are detrimental to the problem-solving process. Fear of personal consequences (which, as discussed, are often real), and lack of confidence leads to retreat, preventing people from speaking up during the investigation. In response to the perceived (or real) threats created by the problem, people were observed employing manipulation, aggression, coercion, and blame. For example, the engineering leader may know that his department has caused the issue to occur, but points to manufacturing for the sake of protecting himself and his function. In aggregate these common themes of retreat, coercion, manipulation, deception, blaming others, and aggression led to the identification of Concept 2, the interpersonal response of people to severe problems. The research shows that, like the Pressure Response, the People Response leads to the breakdown of teams and the problem-solving process, leaving individuals to act independently.

One example of how this happens is that people use the 8D tool to 'point fingers'. One person described how people had 'weaponised the 8D' to shift responsibility. An example to illustrate comes from a business CEO:

> What I have not seen work is, a templated approach to problem-solving. I've not seen that work yet. Where someone comes with the slide deck that says, 'OK, we have to fill out this slide deck and if we fill up the slide deck, we'll have solved this case'. Because in the end you have people, you have a people problem with these types of problems, right? ..... You've got a huge people element which can't be contained there, with these big problems. Which means the behaviour in an investigation is 'who said what, who did what, when did they do it'. So 8D in a big problem, like this is really a witch hunt. It's like a mathematical witch hunt. 8D, it's great for process problems. Terrible as a big hairy problem tool for solving technical problems.

As related by a former managing director for European Operations on how the 8D was used to place blame:

> The environment is unbearable because of the politics and people not being truthful about their input into the problem. It was like, never be caught holding the ball, always make sure it was in someone else's hands. Well for this major problem, we went down an 8D process to try and get to what was the root cause. But that was initially, and then to be quite frank with you, it moved away because the 8D was just people, like it was a tool for people to hide behind, to say 'not me, not me'.

It was reported that in severe event situations, people would abandon traditional approaches or use them incorrectly, substituting their own approaches to resolve the problem. When the formal problem-solving process was abandoned, teams are left unsupported and in 'firefighting' mode. Respondents stated that instead of a method to support the team in identifying root cause, the problem-solving process is used superficially, as a post-crisis framework to produce reports and make PowerPoints. Rather than guide the team in their problem-solving efforts, the process is just a repository mechanism. An engineering director described how the 8D process was abandoned and used only to present a desirable outcome:

> Obviously, we have 8D problem-solving processes which should be, it should have been actually initiated from the very beginning and we haven't done it. We use it more like a slide collection place after you've done everything.

The interview participants described how, when encountering a technical problem that isn't solved by logic, there is an inability for people to make sense of the situation and a loss of faith in the team and the problem-solving process. The workers built the product according to the standard work, the engineers are all fully degreed and experienced, the suppliers are qualified, so how can this problem exist? As related:

> You think you've taken a purely scientific approach, but it doesn't work. You can go back to the people. You can pull the data and it doesn't matter if you don't believe it, don't trust the data. I mean, in your perception, it's not real, it must be incorrect, or someone must have gotten the story incorrect.

The 8D is scientific and a tool that requires rationality, but the interviews indicated a departure from rationality, where interpersonal relations and emotion takes centre stage:

> In a multicultural, global, multinational company when you have experts getting together for the investigation, there's sometimes no sense of the other person's background. The rationale is given, and if it doesn't fit with the expert's understanding, it gets emotional very quick. So, the first meeting with the team to talk about this, turned into a shouting match between experts with over 35 years in the company from different sides of the ocean. And all with the same facts. How do you explain this?

As one interviewee said about dealing with interpersonal relationships in the context of problems, 'Yeah, it's really not easy. I think it's the most difficult thing. The most difficult thing'. So, with this background of pressure on the individuals, not being able to make sense of the situation, and abandonment of the mandated problem-solving method, the teams experience interpersonal People problems: they blame each other, manipulate the process, act with aggression, freeze, panic, and try to escape. From an interview with an Engineering Director:

> People tend to feel like.... shit, 'I hope it's not me'. So, let's just push it somewhere else and someone else will have to prove it is not him, before it's going to get me... I might be lucky then and why not, it doesn't go back to me at all. ... We tend to blame people who are directing and take consequences rather than improving the process around the problem and just treating that actually as a lessons learned. We would actually go after the people.

Instead of the team working on the investigation and root cause analysis, social dynamics break down. Colleagues become a way to escape the pressure. A Quality Leader described this environment of performing the 8D root cause analysis for scrapped components produced for nuclear power applications:

> Sometimes I had big troubles with the production leaders. They feel attacked with a root cause analysis, them personally. And they always tried to move attention to a different focus rather than really collaborating and not in a manner that solves the problem. Move it to somebody else or to another topic. Trying to deviate the attention of the team into another problem which is not the one which is under analysis. We lose time and they make the problem grow. I guess because they're afraid or they think there's going to be punishment or something negative, they feel they have to blame others.

Figure 6 illustrates how the raw data from these interviews evolved into second order themes and terms which, through qualitative analysis, led to the aggregate dimension that is the social response of people and how they behave relationally while problem-solving during a crisis.

Figure 6 Aggregation of first and second order concepts and themes: People[33]

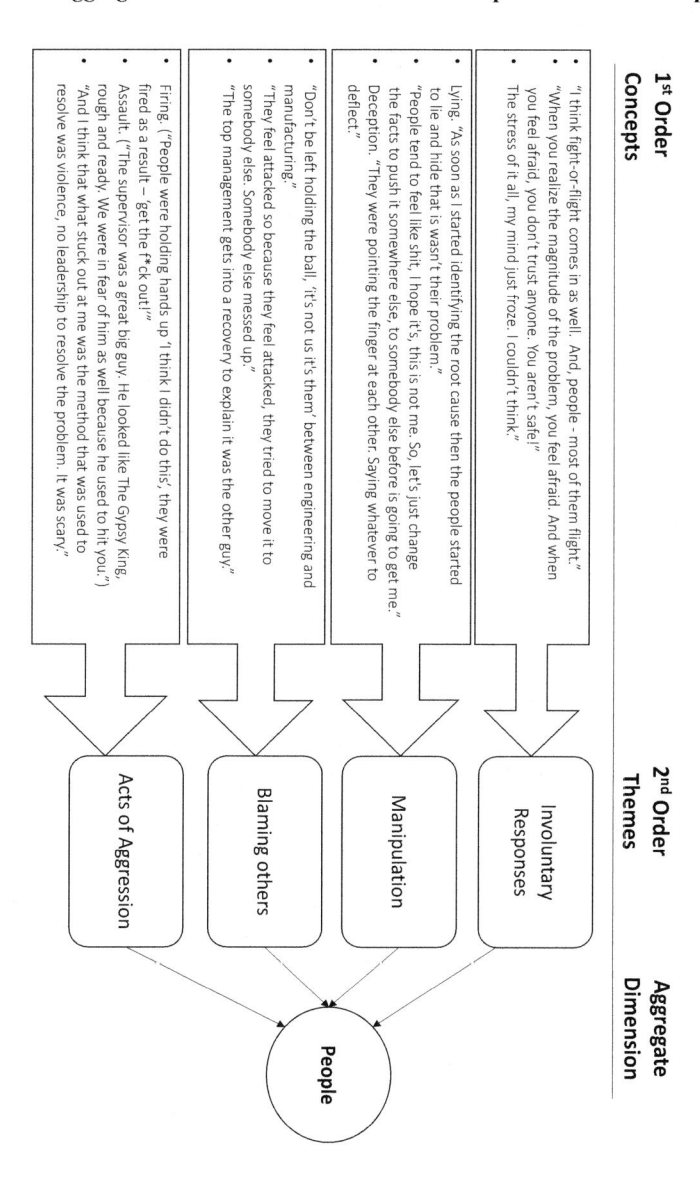

1st Order Concepts	2nd Order Themes	Aggregate Dimension
• Lying. "As soon as I started identifying the root cause then the people started to lie and hide that is wasn't their problem." • "I think fight-or-flight comes in as well. And, people - most of them flight." • "When you realize the magnitude of the problem, you feel afraid. And when you feel afraid, you don't trust anyone. You aren't safe!" • The stress of it all, my mind just froze. I couldn't think."	Involuntary Responses	People
• "People tend to feel like shit, I hope it's, this is not me. So, let's just change the facts to push it somewhere else, to somebody else before is going to get me." Deception. "They were pointing the finger at each other. Saying whatever to deflect."	Manipulation	
• "Don't be left holding the ball, 'it's not us it's them' between engineering and manufacturing." • "They feel attacked so because they feel attacked, they tried to move it to somebody else. Somebody else messed up." • "The top management gets into a recovery to explain it was the other guy."	Blaming others	
• Firing. ("People were holding hands up 'I think I didn't do this', they were fired as a result – 'get the f*ck out!'" • Assault. ("The supervisor was a great big guy. He looked like The Gypsy King, rough and ready. We were in fear of him as well because he used to hit you.") • "And I think that what stuck out at me was the method that was used to resolve was violence, no leadership to resolve the problem. It was scary."	Acts of Aggression	

---

[33] The specific identify of the Gypsy King referred to in the interview was not clarified further, but I believe this to be a reference to the incomparable Tyson Fury, two-time world heavyweight champion, who won me free wings and beer when he defeated Deontay Wilder in 2021.

Discussion: what People problems mean to problem-solving team members

These interpersonal 'People' problems have a deleterious effect on an investigation team. As a corporate lean director communicated during an interview:

> What I see, sorry to say, on big impactful problems, is a lot of finger-pointing and a lack of a sense of ownership. It's a lot of trying to point fingers or make the responsibility vague, instead of taking ownership, and this is very sad, and a huge waste of time because the problem is still there. And with everyone focused on shifting blame, the impact of the problem only gets bigger.

In a problem environment, the individual feels pressure, which then translates into self-protection activities, damaging interpersonal relationships and inhibiting the teamwork needed for a complex root cause investigation. Further, the conditions—information is lacking but answers are required—exacerbate the pressure on individuals, leading to a focus on interpersonal relations and away from the problem itself.

The following section explains that as individuals and the team involved in the problem-solving process lose the ability to function, the process is further impaired by external organisational actors, that is, political forces from outside the team interfering with the problem-solving investigation.

Concept 3: Social Response to Politics

Problems can spread beyond the team employed to resolve them, entrenching management and the larger organisation. As people's responses to intense pressure and fear divide the problem-solving team, leaders and other groups may use strategies and tactics to influence outcomes and shift blame. Evidence was provided of management using influence to place blame on the team, and using power to derail the root cause analysis and even to change and destroy evidence. This phenomenon of responding to problems by employing political influence is particularly relevant in large corporations with matrix organisations, where more than one leader is in charge of the same scope.

When queried about problem-solving severe events, interview subjects raised common themes of management executives using their power to influence the investigation: protecting themselves, changing the results of the documented root cause, and shifting blame and attention to other executives and leaders. In aggregate these practices represent a set of political dimensions that undermine successful problem-solving: the social response to Politics. As one Danish CEO related, when involved in severe warranty events, the political dimensions were often more challenging than the technical problem itself:

> There's been many examples where the technical part might actually be less complex than the political part. Unfortunately, I have been there heading the technical investigation of a couple of events with fatalities. And that is where you know the political agenda might conflict with the technical truth. In this type of an investigation in a multinational company, there are so many functional lines and disciplinary lines. It becomes a very blurry picture: who actually should do what and who actually failed in a situation like that. So that would be too complex to firmly investigate. It became so political that I could immediately feel my support from the organisation disappear. I was on my own.

These political dimensions can derail the problem-solving process. For example, the threat of retaliation can influence the outcome of the investigation. One leader related that when an investigation was leading toward challenging past decisions, leadership would apply the 'Ancient Egyptian method of management': take the investigation report (8D PowerPoint) and 'they take the whip, and just whip you'. In these situations, the 8D method can become weaponised; a 'hate-D':

> You put people in the 8D box, and they do a good job and they get somewhere and they create and they craft right through the process. It gets presented to the senior executive management and because the answers don't culturally fit, or they expose that group to any kind of criticism, then they'll want to influence and change the 8D output. And, by doing so, it completely dilutes the intent and purpose. Then two things happen: one is they destroy the end results of the 8D, or at least pervert it, and second, the group [problem-solving team] then know that their job is to craft something that fits the needs of those people - not necessarily even the business or the customers.
>
> The business shouldn't want them to do that either. They should want a pure solution. So, the 8D is corruptible… a diversionary tactic, a diversionary document you can use to point fingers. We've got a hate-D, and why not?

Because these are severe events with major impact, visible in the organisation and potentially the media, managers use what tools they have available to spread the blame and not shoulder it alone. As related by a Senior Leader of Quality, Health, and Safety:

> If the root cause points at you, says you need to pay them a billion euro and replace a fleet of turbines, you try to fight and

say 'ah that can't be, it's not my entire fault. Somebody else was involved and I'll be sure that this is the right root cause'. You use your influence and ask questions: 'And have you, why haven't you investigated this or why have you made these calculations? Are you sure that it's not this other thing?' So, when the final decision is to be made of who is to pay because that's more or less always what we talk about, who is to pay for this mistake, then no one puts up their hands and says, 'I'll pay for it'. No one does. And you make sure it's not you. Because the root cause points in a direction they don't like. So, they try to deny it and try to manipulate. The different bosses fight to get the outcome of the problem-solving that they want.

Using power to influence and control the narrative was not limited to coercing the 8D or Six Sigma leader to modify the investigation. Examples were provided of executive leaders modifying the operational definition and criteria of 'severe event' to prevent problems from reaching the board of the company:

> The severe event is triggered automatically when the cost is loaded in the system, and the impact is loaded. This is triggered through a tool called 'Service Now'. This is well known and understood.
>
> We had a $500,000 threshold, but because of the number of issues we were seeing within a product, and the continued escalations to the board, we were told to increase the threshold. This is why we had to increase the threshold to a million bucks.

It is relevant to note, while the leaders can be successful in changing or stalling the investigation outcomes, the problems often are not solved and continue to escalate. Multiple respondents noted that they lost faith in and respect for their leaders because of their political handling for personal gain. A Senior Leader of Health and Safety reported being ordered to stand-down and cancel an investigation once the facts revealed mistakes made in a leader's sphere of influence:

> There was a critical fault on the product during the final test before release for shipment; one of the blades came off. I was asked to investigate why the failure occurred, as an independent person. I wanted to do it professionally without being biased. Be, you know, completely open minded about it. So, I started off by gathering evidence and I did that by looking at the project documentation and in doing that I found that there were some discrepancies in the sign-off signatures. And the deeper I got into it the more I found that there had been some sort of fraud going on in the respect of forged signatures.
>
> I presented my initial findings to the Managing Director. After this meeting, I received a call from Human Resources that the interviews were cancelled, and I was told the investigation was suspended.
>
> That's why I sometimes feel that when we do 8Ds we don't really get to the truth. Some very clever people that know all about 8D and the types of Root Cause Analysis findings and problem-solving. It's great to see on the screen, but does it really solve the problem? The Managing Director, so he taught me one thing. The company talks about integrity and respect, but I feel the integrity has been breached in this occasion. And I felt let down. They've lost my trust.

Figure 7 provides a representation of how raw information evolved into concepts and themes through qualitative analysis, which led in aggregate to the third social mechanism identified by this research: Politics.

Figure 7 Aggregation of first and second order concepts and themes: Politics

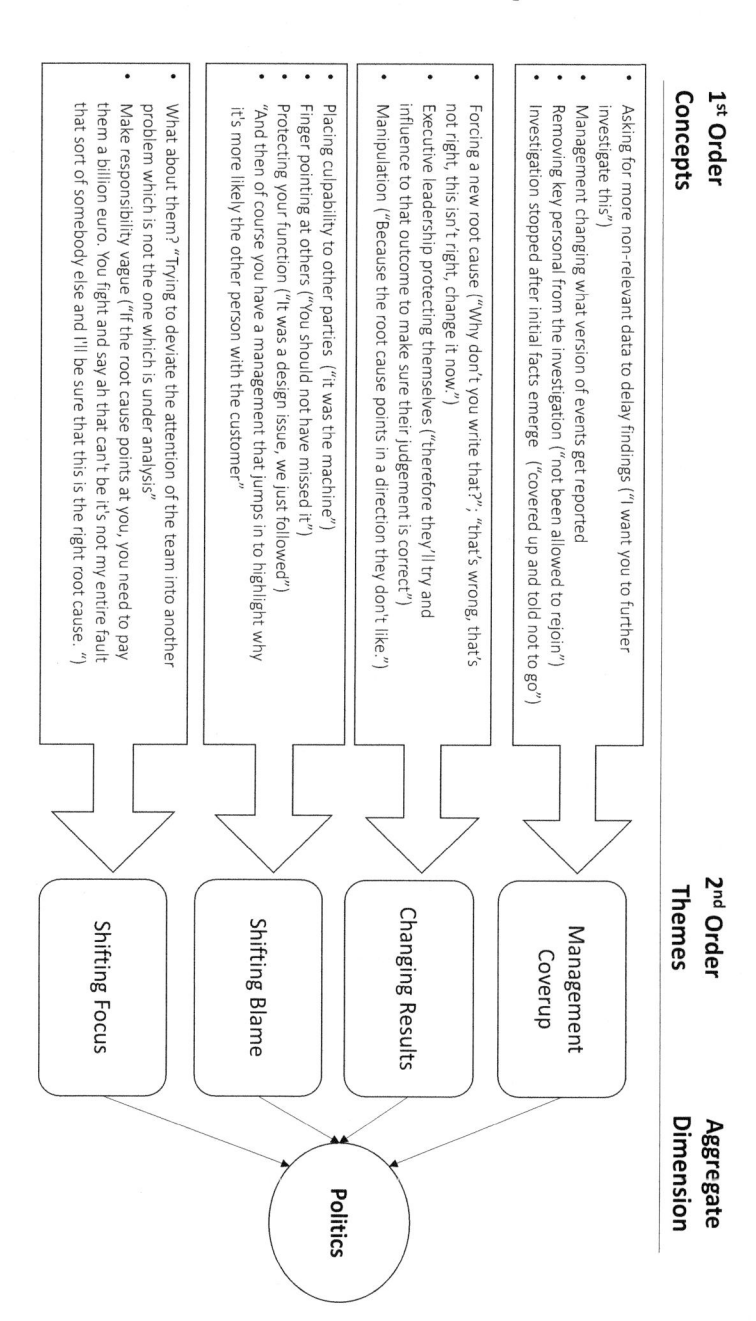

1st Order Concepts

- Asking for more non-relevant data to delay findings ("I want you to further investigate this")
- Management changing what version of events get reported
- Removing key personal from the investigation ("not been allowed to rejoin")
- Investigation stopped after initial facts emerge ("covered up and told not to go")

→ **Management Coverup**

- Forcing a new root cause ("Why don't you write that?", "that's not right, this isn't right, change it now")
- Executive leadership protecting themselves ("therefore they'll try and influence to that outcome to make sure their judgement is correct")
- Manipulation ("Because the root cause points in a direction they don't like")

→ **Changing Results**

- Placing culpability to other parties ("it was the machine")
- Finger pointing at others ("You should not have missed it")
- Protecting your function ("It was a design issue, we just followed")
- "And then of course you have a management that jumps in to highlight why it's more likely the other person with the customer"

→ **Shifting Blame**

- What about them? "Trying to deviate the attention of the team into another problem which is not the one which is under analysis"
- Make responsibility vague ("If the root cause points at you, you need to pay them a billion euro. You fight and say ah that can't be it's not my entire fault that sort of somebody else and I'll be sure that this is the right root cause.")

→ **Shifting Focus**

2nd Order Themes → **Aggregate Dimension: Politics**

Discussion: What political dimensions mean to problem-solving team members

The effect of political influence prevents successful problem-solving during the investigation and traditional problem-solving efforts are perverted, misdirected, and thwarted by managers. As related by a Senior Leader of Quality, Health, and Safety: 'For the root cause analysis.... the problem is we do not have anything to control the political game around this, around the problem itself. And that is missing and the 8D or the Six Sigma or the CAPA or the whatever we are using is not sufficient. The methods don't help'.

Chapter conclusion and next steps

The 3P effect undermines the linear decision-making process required for a traditional technical problem-solving approach to be successful. It makes it more difficult to make sense of what is happening, eroding people's ability to respond constructively and sometimes even exacerbating the problem itself. These social mechanisms of Pressure, People, and Politics hijack and sabotage the problem-solving process under certain conditions. The next chapter identifies the conditions that are present when 3P responses are triggered, the S.H.I.T. problem. The identification of a S.H.I.T. problem is the first step to empower companies to navigate and solve severe technical problems faster and more effectively, and to avoid both economic and human costs.

CHAPTER 3 When Traditional Problem-Solving Fails: S.H.I.T.

Sometimes, perhaps even most of the time, the traditional approach to problem-solving works. But sometimes it doesn't. As described in the previous chapter, expert interviews identifying actual examples where traditional problem solving failed to resolve a severe problem were analysed using the Gioia approach. First-order coding presents data (verbatim) from the interviewees while second-order analysis organises the data under broad concepts and themes. Though the type of problems varied, certain circumstances occurred again and again when traditional systems failed. These common characteristics were described repeatedly from all areas around the world and in multiple business models. They are:

- **Severity**: The technical problem has a significant impact on the organisation.
- **High visibility**: The problem is highly visible internally or externally.
- **Information deficiency**: Information needed to solve the problem is not available.
- **Time is sensitive**: The negative impacts of the problem are continuing and/or worsening with the passage of time.

These problem characteristics have been combined into a mnemonic framework that can be used to signal that the traditional problem-solving approaches will be inadequate: a S.H.I.T. problem. Figures 8 through 11 illustrate the emergent themes around contextual conditions present when traditional problem-solving approaches fail in the face of severe problems, showing the process for how the S.H.I.T. framework was developed.

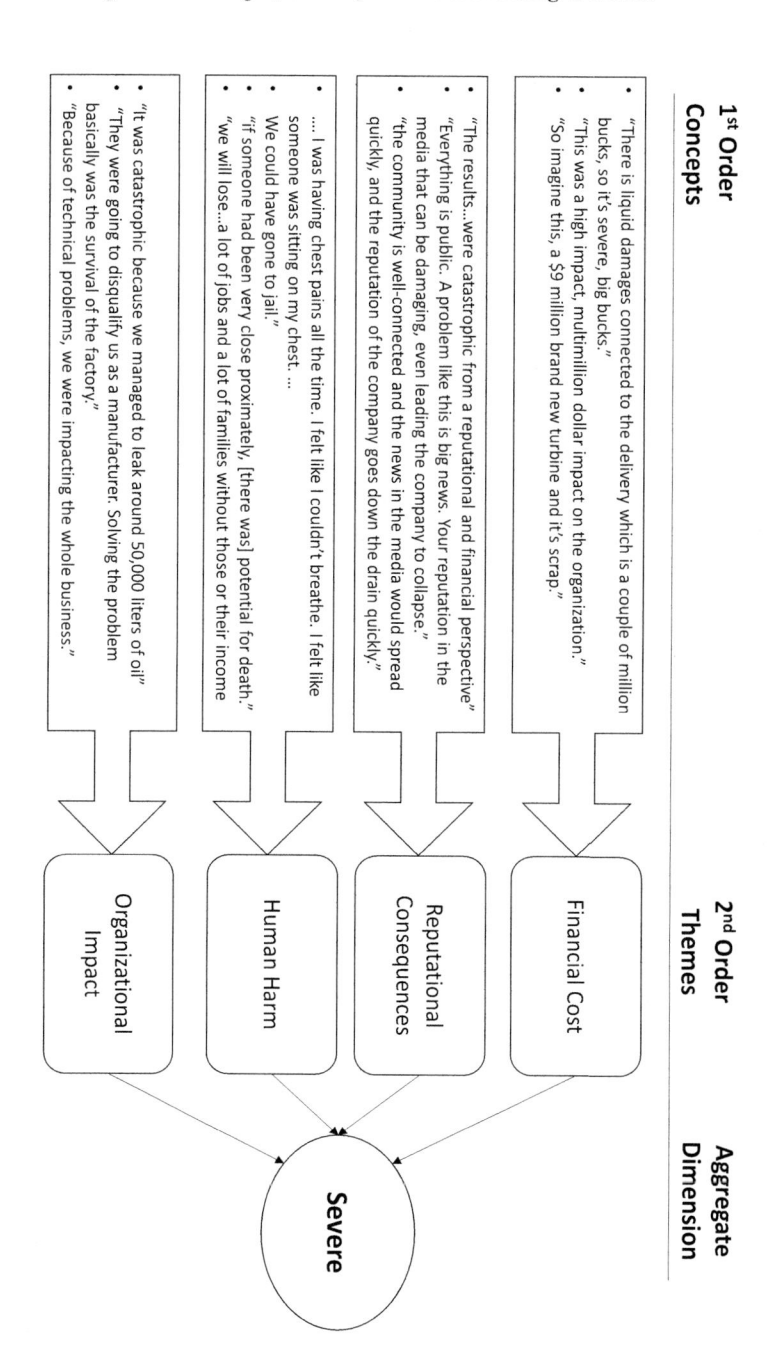

Figure 8 Concepts, themes, and factors leading to Severe

1st Order Concepts	2nd Order Themes	Aggregate Dimension
• "There is liquid damages connected to the delivery which is a couple of million bucks, so it's severe, big bucks." • "This was a high impact, multimillion dollar impact on the organization." • "So imagine this, a $9 million brand new turbine and it's scrap."	Financial Cost	
• "The results...were catastrophic from a reputational and financial perspective" • "Everything is public. A problem like this is big news. Your reputation in the media that can be damaging, even leading the company to collapse." • "the community is well-connected and the news in the media would spread quickly, and the reputation of the company goes down the drain quickly."	Reputational Consequences	Severe
• ....I was having chest pains all the time. I felt like I couldn't breathe. I felt like someone was sitting on my chest. ... • We could have gone to jail." • "if someone had been very close proximately, [there was] potential for death." • "we will lose...a lot of jobs and a lot of families without those or their income	Human Harm	
• "It was catastrophic because we managed to leak around 50,000 liters of oil" • "They were going to disqualify us as a manufacturer. Solving the problem basically was the survival of the factory." • "Because of technical problems, we were impacting the whole business."	Organizational Impact	

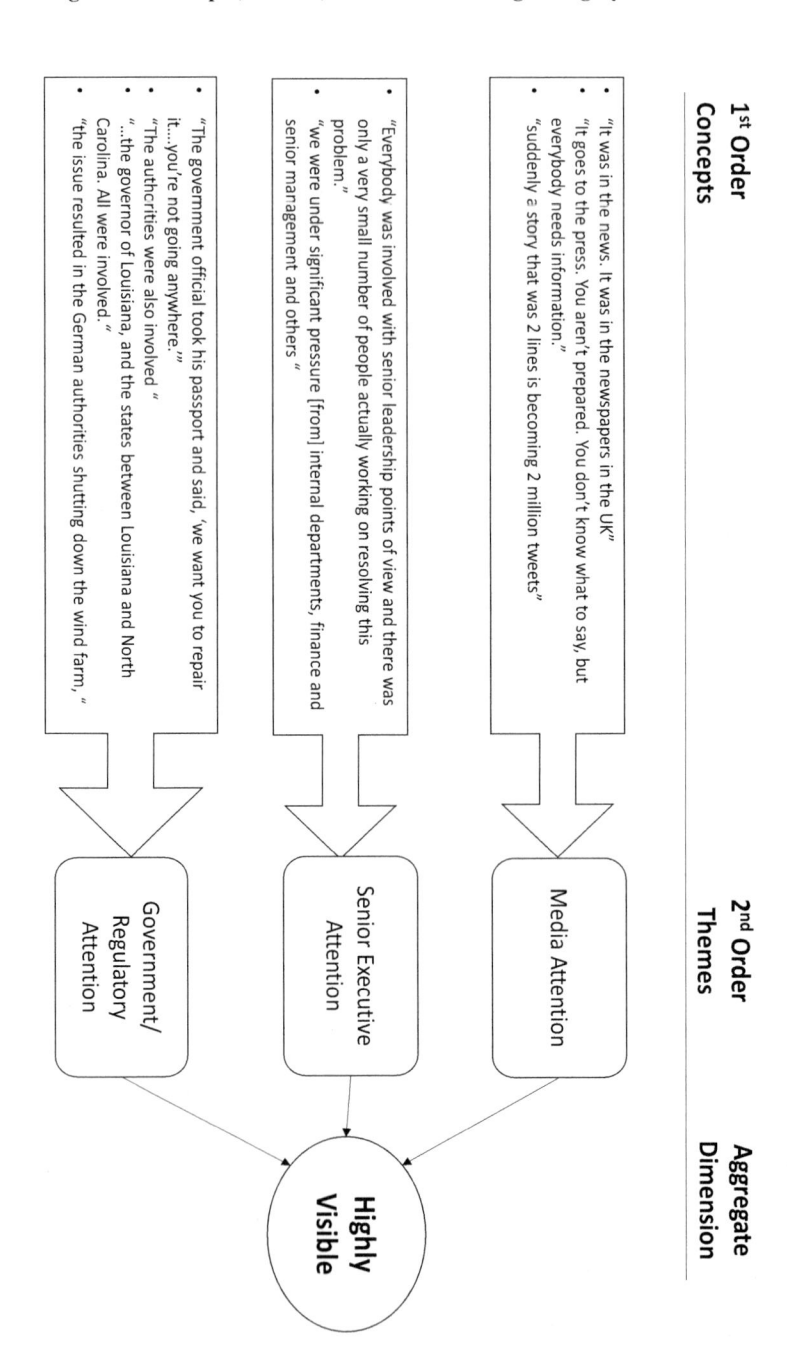

Figure 9 Concepts, themes, and factors leading to Highly Visible

1st Order Concepts	2nd Order Themes	Aggregate Dimension

1st Order Concepts:

- "It was in the news. It was in the newspapers in the UK"
- "It goes to the press. You aren't prepared. You don't know what to say, but everybody needs information."
- "suddenly a story that was 2 lines is becoming 2 million tweets"

→ **Media Attention**

- "Everybody was involved with senior leadership points of view and there was only a very small number of people actually working on resolving this problem."
- "we were under significant pressure [from] internal departments, finance and senior management and others"

→ **Senior Executive Attention**

- "The government official took his passport and said, 'we want you to repair it….you're not going anywhere.'"
- "The authorities were also involved."
- "…the governor of Louisiana, and the states between Louisiana and North Carolina. All were involved."
- "the issue resulted in the German authorities shutting down the wind farm,"

→ **Government/Regulatory Attention**

→ **Highly Visible**

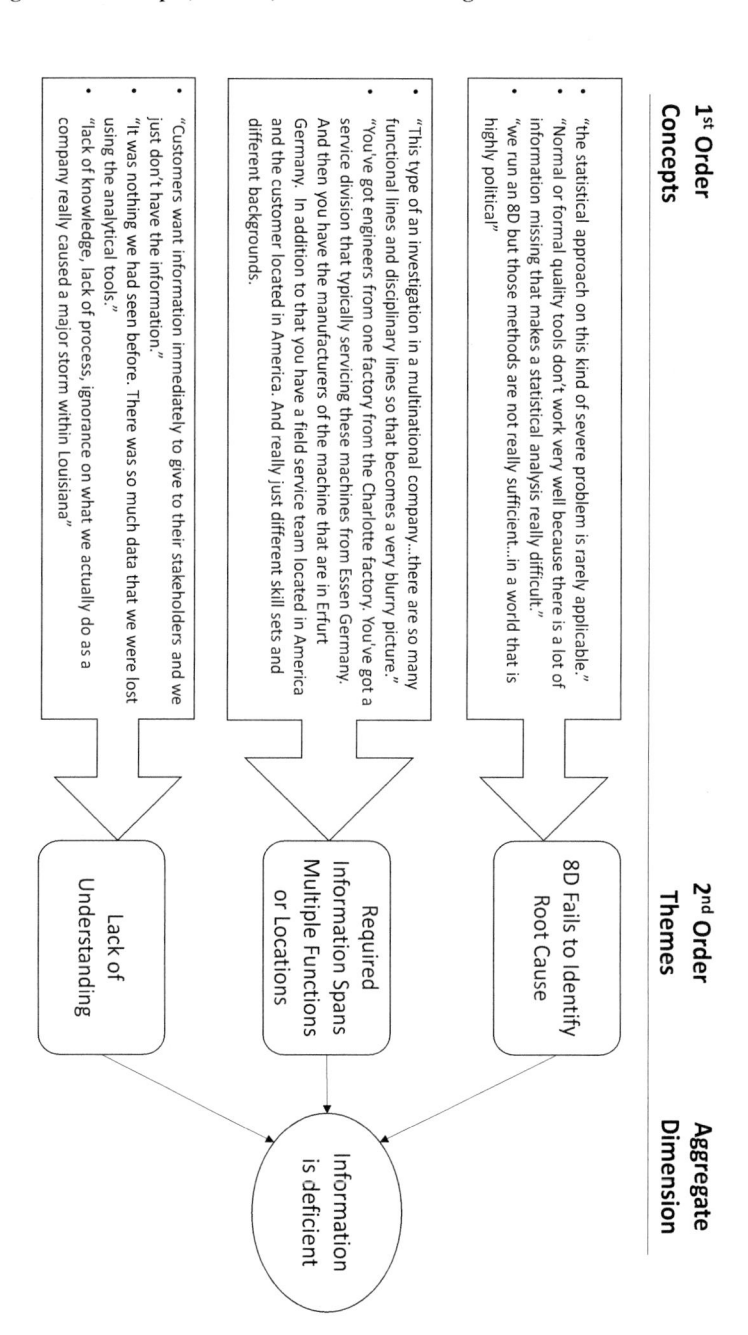

Figure 10 Concepts, themes, and factors leading to Information deficient

1st Order Concepts	2nd Order Themes	Aggregate Dimension

- "the statistical approach on this kind of severe problem is rarely applicable."
- "Normal or formal quality tools don't work very well because there is a lot of information missing that makes a statistical analysis really difficult."
- "we run an 8D but those methods are not really sufficient...in a world that is highly political"

→ **8D Fails to Identify Root Cause**

- "This type of an investigation in a multinational company...there are so many functional lines and disciplinary lines so that becomes a very blurry picture."
- "You've got engineers from one factory from the Charlotte factory. You've got a service division that typically servicing these machines from Essen Germany. And then you have the manufacturers of the machine that are in Erfurt Germany. In addition to that you have a field service team located in America and the customer located in America. And really just different skill sets and different backgrounds."

→ **Required Information Spans Multiple Functions or Locations**

- "Customers want information immediately to give to their stakeholders and we just don't have the information."
- "It was nothing we had seen before. There was so much data that we were lost using the analytical tools."
- "lack of knowledge, lack of process, ignorance on what we actually do as a company really caused a major storm within Louisiana"

→ **Lack of Understanding**

→ **Information is deficient**

Figure 11 Concepts, themes, and factors leading to Time-Sensitive

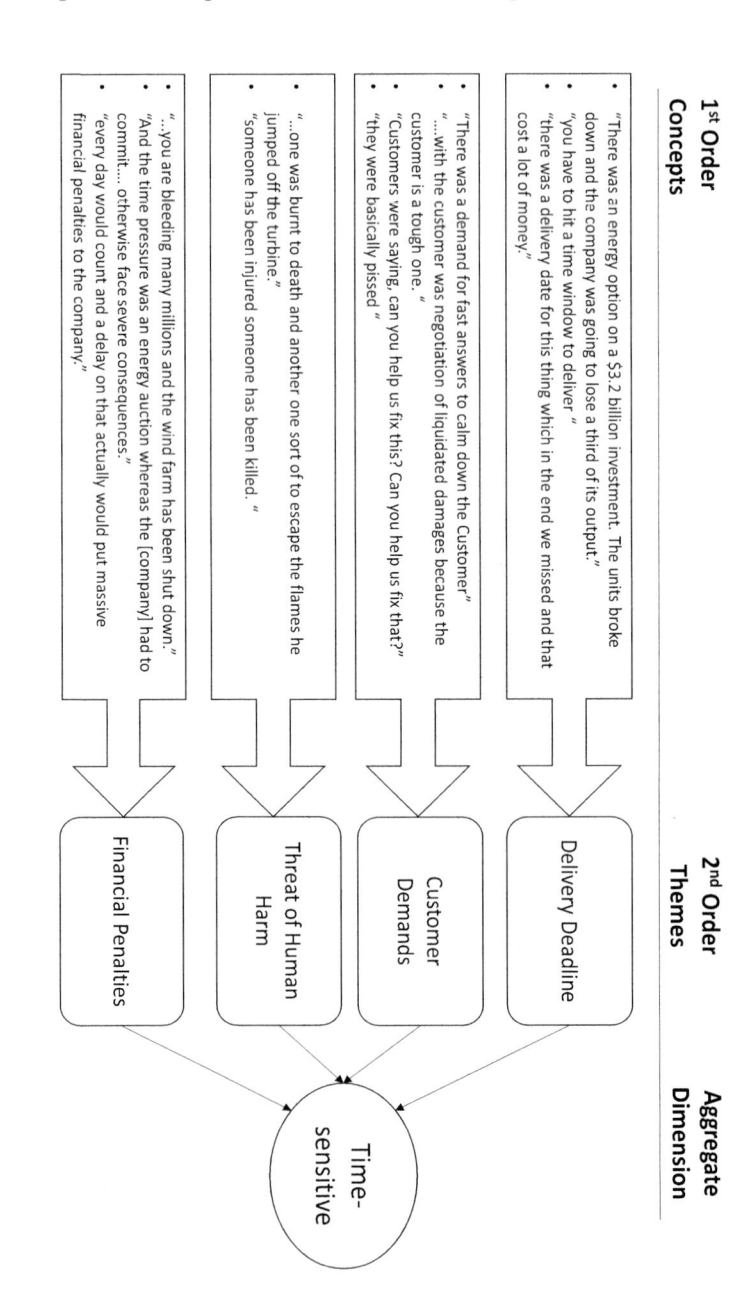

1st Order Concepts	2nd Order Themes	Aggregate Dimension
• "There was an energy option on a $3.2 billion investment. The units broke down and the company was going to lose a third of its output." • "you have to hit a time window to deliver." • "there was a delivery date for this thing which in the end we missed and that cost a lot of money."	Delivery Deadline	
• "There was a demand for fast answers to calm down the Customer" • "...with the customer was negotiation of liquidated damages because the customer is a tough one." • "Customers were saying, can you help us fix this? Can you help us fix that?" • "they were basically pissed "	Customer Demands	Time-sensitive
• "...one was burnt to death and another one sort of to escape the flames he jumped off the turbine." • "someone has been injured someone has been killed. "	Threat of Human Harm	
• "...you are bleeding many millions and the wind farm has been shut down." • "And the time pressure was an energy auction whereas the [company] had to commit..... otherwise face severe consequences." • "every day would count and a delay on that actually would put massive financial penalties to the company."	Financial Penalties	

Industry mandates use of specific problem-solving techniques that are linear and deductive. And for many technical problems these techniques are highly effective. However, the case study research revealed that when S.H.I.T. conditions are present, people do not respond in a rational manner and linear problem-solving techniques faulter. The S.H.I.T. framework identifies the problem context, the conditions surrounding a problem, that signals that traditional problem-solving methods will not be enough to successfully resolve the problem. This set of conditions was identified using analytical tools established in academic research on behavioural science. Rather than repeat that here, and since an easier way for most people to learn and understand concepts is through stories, I relied on 27 stories to identify the S.H.I.T. framework. I have included just three here. These three stories illustrate S.H.I.T. using real-world problems.

Even though you know what to look for, as you navigate a high-stress environment and a rapidly evolving set of facts, identifying when you're in the middle of S.H.I.T. isn't always easy. In each story I encourage readers to identify the first time they recognise one of the four S.H.I.T. conditions. Hint: All three stories have all four conditions. If you want to be an overachiever, when you get to the 'S.H.I.T. Problem-Solving Identification Evaluation', a tool for identifying and reporting the presence of a S.H.I.T. problem, come back to these stories and use the tool to document your findings.

Case 1: Product Design Failure led to Power Blackouts in Africa

The Background

The information included in Case 1 came from interviews and descriptions provided by three senior leaders working in the power generation industry: the Quality Leader; the Managing Director responsible for production and manufacturing at the plant, and the Product Line Director managing Sales and Engineering. The decision to focus the interview on this specific problem came directly from the interviewees independently of each other, as a signal of the enduring scope and impact this failure had on the organisation.

The focal incident relates to a generator alternating rotor, designed to spin over 3,500 cycles per minute, which was ordered, new, from a 100-year-old U.K. plant that had the ability to design, produce, and service these electrical engineering products. The plant producing the part belonged to an American-owned engineering corporation with over $100 billion in annual revenue.

A generator rotor is a highly engineered, customised component with a typical 18-month lead time from order to customer delivery. Due to the critical nature of the component in the power station, before the product is released for shipment, it is fully tested electrically and mechanically to validate the component for safe operation.

The order came from a power station operating in South Africa, commissioned in the 1980's and fully operational by 1990. With almost 30-years of service life, the current rotor needed replacement. The customer operating this coal-fired power station had running installed capacity of approximately 4 gigawatts which served as an important element to the South African energy infrastructure. A power station of this size is equivalent to 1,524 onshore wind turbines or 12.5 million solar panels[34]. The energy produced from this generator powers over 2.4 million households in the African state.

Characteristics of the Problem

A description of the problem from the Managing Director of the plant at the time of the event:

> Yeah, so this, the rotor, brand-new rotor, we wound the rotor, so we put the copper coils in we then put it into the test facility to then test and balance the rotor, **and it failed!** [Emphasis by the interviewee]. And it failed because on a generator rotor there's obviously a lot of insulation, a lot of high voltages, and the installation failed, and it cracked and was split into pieces. Major non-conformance! [Interview – Managing Director, Plant]

The failure of this component and its severity was immediately recognised by the customer. The emphasis was placed on the potential for high impact:

> So, we were right by the delivery date when it failed with a very sensitive customer who needed the rotor. So, it wasn't just they were missing a delivery date, they actually, needed the rotor for power generation in South Africa! Holy shit, South Africa was having rolling blackouts at the time and we were getting the blame by the customer. [Interview – Managing Director, Plant]

---

[34] Source: USA Department of Energy
https://www.energy.gov/eere/articles/how-much-power-1-gigawatt

The bold headline published in a major international business periodical article noted the disappointment to the African people. Respondents recalled that a formal 8D technical problem-solving investigation was immediately started to understand the root cause of the failure and generate improvements, but it was subsequently manipulated for the purpose of assigning blame to other departments instead of promoting team-work and collaborative problem-solving required for a rapid problem resolution.

> So, we pulled together an 8D team, you go through, and you try and do a cross-functional team. So, you get the engineers involved and you get the guys who've built it [production operators] and done that, and again you're going through and you're using techniques like Ishikawa [problem solving tool], you try and do the 'five whys' again, why would something fail? And we got to the point where we actually were saying, well actually we can't find any reason from the manufacturing point of view as to why this would have failed. So, we stripped it and we replaced the insulation that had cracked and we went through and we said, right okay, why would the insulation crack? And all the reasons came back from the Engineers was that it must be poor manufacturing…. engineering, they're just shooting at you…placing the blame, 'Yeah, it's manufacturing, it wasn't assembled correctly and there could be no other reason why the insulation cracked other than poor manufacturing'. And even the head of engineering globally, for generator rotors, even actually said that on a call. 'You know, we've looked at everything, we can't see why this would happen so it's got to be a manufacturing issue'. They told us, it was poor manufacturing, it had been fitted incorrectly, it had been assembled incorrectly. [Interview – Managing Director, Plant]

Without a definite root cause on the source of the technical issue, but with pressure and daily financial penalties that were 'in the tens of thousands of euro a day from the customer due to the late contractual delivery' (Quality Leader), the team decided to repair, rebuild, and retest the product. The team reassembled the rotor and proceeded again for the final testing and validation... which then failed a second time.

The organisation then relaunched the 8D problem-solving investigation. The Managing Director explains the environment and the effectiveness of the 8D on problem-solving:

> Well, we went down again the 8D process to really, you know, try and get to what was the root cause. But, to be quite frank with you, it moved away from the 8D. It was a tool for people to hide behind if you like, to say 'not me, not me! I did everything I could, must be someone else'.
> Yeah, that was quite bad really. People running scared.
> And so, all the way through this, and this comes back to if there's an issue, the manufacturing they're always left with the ball. They're always left with the problem because the customer wants it. And the more you squeeze on that it's more about what can manufacturing do to reduce their lead time. So that introduces new dynamics, like you could be introducing problems where there [aren't] problems because of time pressure, because people start taking shortcuts, so it can give you false readings. So, the time pressure creates problems there, but they're not real, they're only problems created by that scenario rather than the actual root cause on this job, rather than there was a design issue. There's a fundamental design issue. That's [the] best way to put it, it's a symptomatic problem from the environment that is now finding itself in, which is time pressure, the politics. [Interview – Managing Director, Plant]

The component, urgently needed for power generation in South Africa, had failed its quality and safety tests twice. External media reports of blackouts in the destination country were drawing attention. Internally, after the second failure, the customer demanded that the CEO of the global company come to South Africa to present the recovery plan personally. Amidst this, the restarted 8D was then abandoned by the teams once they realised how severe the technical issue was and the investigation was completed independently by the engineering function.

Summary of Problem Impact

The total economic impact of the rotor rewind technical failure was over USD 3 million: USD 2.6 million from liquidated damages to the customer as a penalty for not producing electricity and over USD 600,000 from labour costs to correct the non-conformance in the plant. However, the outcome included not just economic loss, but multiple failed attempts to resolve the issue due in some part to social factors personally impacting the stakeholders of this event.

The Effect of Human Behaviour on the Problem-Solving Process

As the Managing Director said, experiencing these types of problems feels like 'the world's gonna end…' The intrinsic properties of these big problems can hurt the 8D team members.

The Quality Leader at the time described these factors and the event environment for the working team:

> The problem had time constraints needing to come to a quick solution and outcome due to the outage … very emotional

> customers at the end of it… the end customer was very animated and used foul and abusive language in some of the reviews. Accusing us of all sorts of things from sabotage to lying to doing things underhandedly, forging paperwork. [Interview –Senior Quality Leader]

Speaking of the strain to complete the job, he noted how the severe event added additional business pressure at multiple levels:

> So, there was a massive financial pressure to ensure we got the contract completed to recognise revenue, 'rev rec' [the financial position], so we could say we got the money complete and in the business for the month and quarter end.
>
> We had internal customers, let's say, different part[s] of the business that placed the order upon the plant, writing all different levels across the business, top and down, that if we did not complete on time the customer would cancel the order: 'This customer was a really good customer, we have a long-term contract with them, you are letting us down if you don't achieve this, [the company] is going to be seen in a really bad light…'
>
> This was being driven at us daily, hourly at times. You must hit this date or else, to a point where the executive leadership team actually became the supervisors of the job and therefore came in on dayshift and nightshift to make sure we did deliver this X-works per the request from everyone above… Rather than determining the root cause, some senior management from corporate would go 'that is the root cause, that is what you are going to come up with in your investigation'. But you have to allow it [the problem-solving] to go through its course. And we weren't given the time. [Interview –Senior Quality Leader]

The Quality Leader describes the personal and emotional impact of working with this:

> You take it personally. You think you are not doing things right, are not doing things correct[ly]. Especially when you have reviews with senior leadership team, you end up going to a review for an hour and get chewed [out] on every single word or spelling mistake. And rather than agree or challenge in a supportive way, a lot of the time you come out of those reviews feeling completely deflated, dejected. Questioning your own skill, questioning the team skill, have I been going [in] the wrong directions? A lot of the time it's, 'that's wrong, that's not right, this isn't right, this doesn't reflect this'. So, you come away and just feel just completely dejected and think 'I thought I was quite good at this. I am not anymore'. You really take it personally that you are not actually delivering what people expect, or what you think people expect. So therefore, the pressure on yourself starts really building….
>
> I needed to go and see a doctor; it was a case of continu[ing to] drink more alcohol and hoping it would go away or actually think[ing] about now I need to get it sorted. I was having chest pains all the time. I felt like I couldn't breathe. I felt like someone was sitting on my chest. … And in the end, I was signed off work by the doctor. [Interview – Senior Quality Leader]

Case Study Highlights

Characteristics of the problem described:

1. The problem was triggered by a severe technical event: a rotor failed functional and safety tests before dispatch and this failure has the

capacity to harm people and to create significant financial harm to the company.

2. It was a highly publicised and visible in the organisation: external pressure from the customer and in the media, which was exacerbated internally as the team could not find a technical root cause analysis and solution.

3. The problem was time-sensitive due to financial penalties and the urgent need for a new rotor. In this scenario the 8D process failed. It was started (twice) but the process was stopped before a resolution was reached. Only once the event was complete was the 8D retrospectively updated for use as a communication vehicle.

4. The social impact of the problem included a physical toll on the participants where they experienced fear and anxiety over the potential personal and professional consequences of the severe event. These created further strain on the relationship between the functions, resulting in self-preservation instead of solving the technical problem.

Case 2: Steam Turbine Scrapped Component

The Background

Case 2 is focused on a newly manufactured steam turbine component, produced in the U.K. and destined for the North American market. A turbine is the workhorse of modern power generation: in the U.S., for example, turbine technology supports over 80% of power generation[35]. Turbine components are used within nuclear, coal, and natural gas-powered plants, and with different types of biomass applications. Manufacturing accuracy and precision includes the rotor shaft and many ancillary parts such as rotor blades, pins, and other parts. Safety is of paramount importance in the power plant as the risk of a blade failure detachment for a machine operating at 3,600 rotations per minute looms large with failure resulting in a chain reaction potentially destroying the power plant facility.

For Case 2, the interviewees represent Project Management, Engineering, and Health and Safety perspectives. Case 1 illustrated technical problem-solving using 8D when impact could be mitigated by repairing a part. In contrast, Case 2 provides an example when the part is rejected and must be scrapped, losing all value for the company.

Characteristics of the Problem

A description of the problem from the Senior Engineering Leader explains the quality nonconformity and the severe impact of the event on the organisation:

> The U.K. leadership team have all received an e-mail from the plant saying that on [CUSTOMER NAME] particular rotor, we had a massive failure. We have managed to produce a defect with potential[ly] severe customer effect[s]. When we talk about the customer we talk about deliveries and talk about potential financial penalties. This is a massive impact. We're not talking

---

35 SOURCE: US Energy Information Administration
https://www.eia.gov/energyexplained/electricity/electricity-in-the-us.php

about design in production with 20 rotors per year, rather this is a custom-engineered product. This is a massive, colossal impact to the business…

So, we had a situation [where the machine] cut 'a true incorrect path' on the turbine rotor, and managed to create a defect actually on the component. The defect was not repairable. And remember, we're talking about a big component which costs in the millions. And for the customer, it's not all about the money issue for the component of this size and scope; it's about the delivery dates. You spend up to a year to manufacture the turbine [in the factory]. … every single day [of] delay has an impact on your delivery date. And it hurts the customer because they have larger loss of revenue every single day you don't deliver and can't operate the turbine in the power plant. And another impact of this turbine problem is that if you've failed it's not just the time you spent on manufacturing it [the year of factory manufacture], but to get [a] new material that's another six months. Just to get the raw material! … So, delay actually is … you are talking about a year and a half or so, even more than that if there are any more problems. Yeah. So, it's quite tragic.

[Interview – Senior Engineering Leader]

Executive leadership immediately launched an 8D to understand the technical failure and a daily meeting was organised to track status for a product replacement (e.g., 'containment' in the language of 8D). The Senior Leader of Project Management at the time of the incident described the daily meetings and how senior management was using its power to attack the problem-solving team and pressure them to go faster:

We're manufacturing this new rotor [for a new and financially important customer] … And then we start the daily reviews. The executives began hammering everybody to get this rotor to a certain point in time, which is for financial[ly] driven reasons…

> And, that generated a lot of problems and issues. People [were] under great pressure, and then they failed. They failed miserably. They would then apply pressure again, in layers of pressure, both to management and inside the business, so they then transmit that down the organisation. And, they keep pressing and pressing and pressing, and ultimately, then you get to start getting negative reactions and so on, as well.
>
> [Interview – Senior Project Management Leader]

In this case, the 8D problem-solving approach actually increased the pressure and feelings of failure experienced by the team and that high pressure and sense of failure was being transmitted throughout the organisation. When asked a follow-up question on what the role of the 8D technical problem-solving process was in these daily executive calls and how it supported the role of communication, the interviewee reported that:

> I think 8D is a way of moving blame to a different area—very much in terms of deflecting the spotlight off yourself, 'look over there'. And sometimes you can generate a lot of noise in the 8D
> …
>
> But I think in the end, a lot of the stimulus response side for the 8D was from the pressure as well, basically. It's almost like being pushed into an artificially accelerating 8D too, because people [are demanding] 'now, now, now, now…' [Interview – Senior Project Management Leader]

The interviewee observed the 8D process being manipulated by human actors trying to protect themselves. A key element that makes this kind of behaviour possible is that the 8D process, a linear process, requires information to move forward. In this case, critical information about what caused the problem was not available and team leaders took the opportunity to fill in the needed blanks with assumptions that were favourable to them (and assigned blame to others). The 8D process was manipulated and could offer no solution for mitigating these kinds of human behaviours. The Engineering leader described how the deficiency of information needed to solve the technical root cause led to frustration of the 8D process:

> The first assumption, actually from [the] operations team was that this is a machining fault, but how to prove [it]? The assumption was that the machine actually somehow 'freaked out' and decided to cut out of the expected path. I think that was a statement, actually, that particular week and I've been asked to bring someone from the OEM [Original Equipment Manufacturer] or someone who did actually for us [the] retrofit of this machine to come over and explain to us how that happened. So, let's say that it was Monday when I got involved. By Wednesday we had, we had a vendor actually who does all of those retrofits on those machines, and he did the retrofit on this particular machine as well. So, I followed up with him asking can you explain to us how that would happen. What would have to happen with the machine to actually have this defect. His statement was basically based on his experience in industry and based on his knowledge about the industry and I think that's around 20 plus years. Well, he's never heard of a situation like this caused by the machine itself. He could say things goes wrong from time to time but if it happens they don't go wrong by 5 to 10 millimetres, they go horribly wrong which means the blade would actually cut through the rotor rather than just five millimetres. We had no idea of where to go from there. We could

not answer the questions asked. [Interview – Senior Engineering Leader]

The Effect of Human Behaviour on the Problem-Solving Process

The second interviewee in this case study, the Project Management leader, spoke about working on a technical engineering problem-solving team with a problem of this magnitude.

This feedback highlights the pressure on the individual, knowing the visibility of the problem in the organisation and the impact on not solving the problem:

> The driving pressures and the parts [external to] the business can change the culture inside the business significantly. And that's when people… that's when it starts to go wrong. And people get under a lot of personal pressure because they're being… it's almost—don't get me wrong—it's almost like school bullying. It's come down to… Industrial bullying is just as interesting as any other bullying in terms of how it's managed and manipulated. It's very much this… applying to the pressure points. Pressure points might be employment and status. So, I think those kind[s] of pressures put personal pressure on people which makes them change their behaviour…they start transmitting that to other people and you get a domino effect…It becomes a very unreal situation. It's not good and not healthy…

> When you're really being hammered… What are you going to do about this? 'Come on, you know, after two weeks you, still haven't give me anything back on this. What am I paying you for?'…. The person on the other end is not getting any … will

just shut down.... They'll lose all focus on what they should be doing.

And during the 8D, instead of learning on what went wrong, and how we fix it, what it does is, it tends to drive people to say what they believe the protagonist needs to hear, rather than that which is a solution to the problem, and people, people get drawn into a corner. They'll say what they want to; it's a confession under pressure. [Interview – Senior Project Management Leader]

The third interviewee, the leader from Engineering, describes:

All hell was breaking loose. No one cared about solving the problem or [about] the customer. Maybe because people, rather than looking for the cause itself .... were looking to hide and protect themselves or friends. By 'My friends.' I mean let's say, instead of being open and honest and what's the real root cause so we can just fix it and move on, the team was made to feel like shit. [The thinking was] let's just, let's just push it somewhere else [so] that someone else [has] to prove actually it is not him before it's going to get me... I might be lucky and someone else loses their job. [Interview – Senior Engineering Leader]

He further described that an emphasis on placing blame instead of on problem-solving evolved during this event:

People are blamed and this stops follow up actions, actually, with severe events like this...[W]e tend to actually, in advance, blame people who are directing and take consequences [for] people rather than improving the process [around] and just treating that actually as a lesson learned. We would actually go after the people. We would fire them.' [Interview –Senior Engineering Leader]

Case Study Highlights

The problem described in Case 2 encompassed the following characteristics:

1. The problem was triggered by a severe technical event: a rotor had a technical non-conformance and had to be scrapped. Words used to describe the event by the interviewees: 'huge problem', 'critical problem', and 'major problem'.

2. The problem was highly visible in the organisation; senior executives were involved in the investigation.

3. Information needed to determine the root cause was deficient. Some members of the organisation used the information deficiency to manipulate the root cause to point away from the internal organisation to external factors, which frustrated the investigation.

4. The problem was time-sensitive due to financial penalties, the urgent need for a new rotor, and the long lead time of replacement components. In this scenario, 8D was started but the process was manipulated to shift blame. It was only once the event was complete that the 8D was updated retrospectively for use as a communication vehicle.

There is agreement among all three interviewees that human behaviours in a high-pressure situation contributed to a failure of the 8D process. The process was manipulated by people acting in self-interest, which delayed a resolution to the problem, undermined the problem-solving process, and placed a psychological and emotional toll on team members that was transmitted throughout the company. The team did not have the tools they needed to account for, mitigate, or respond to these types of human behaviours.

Case 3: Wind Turbine Service Technician Fatality

Background

A wind turbine is a more complicated machine than many realise. Wind turbines are, in effect, individual single mini-power plants, with the turbine being powered by wind rather than the steam from a boiler. The nacelle, the machine room atop the wind turbine, holds many moving parts to turn wind into energy. The wind turbine needs regular maintenance to continue to operate successfully. With tower heights of up to 153 meters in the air, service operators must work in a high-hazard workplace, replacing and servicing components to prevent and manage natural wear and tear of the parts. These 'turbine cowboys' face intense weather, great heights above the ground, and electrocution on a routine basis.

Three leaders in the wind turbine industry were interviewed about an on-the-job fatality of a wind turbine technician: a former Quality Leader for the Service Division of a major wind turbine producer; a former Product Integrity Leader at the same producer, currently serving as CEO of another wind firm in Denmark, and finally, a Vice-President of EHS (Environmental, Health and Safety) in the wind power industry.

Figure 12 shows a fire in the nacelle of a wind turbine, clearly illustrating the difficulty of escape for workers in the nacelle at the time of a fire.

Figure 12 Example of wind turbine failure mode - fire

Characteristics of the Problem

Two technicians were killed during a fire in a wind turbine nacelle, creating a severe and immediate problem for the entire industry. Here is a description of the problem from the Quality Leader at the time of the event, discussing how it affected their company:

> We have had several issues especially HSE-related [health and safety] issues where someone has been injured, someone has been killed. There is immediately a high-level focus because it will go to the press, and I had several kinds of these that sort of hit us like a lightning striking where we were not prepared. And we needed in a hurry to come up with information control and come up with the right information that we could give to the people. We had there, one I could mention, that started at one site that was in operation, post commissioning and being

> serviced. They had a service crew .... on a turbine and it was so unfortunate that ...this turbine caught on fire while they were up in the top of the turbine, and they had no rescue gear with them. So, it ended really tragically. One [person] was burnt to death and another one....to escape the flames he jumped off the turbine. And it was a massive, huge thing. [T]he authorities, especially authorities in Europe and U.K., the HSEs [health and safety executives] sort of look into how is it actually with the rescue from turbines to ensure we can prevent deaths in the future. That was the CAPA [corrective action, preventive action from the 8D approach] we needed to work on. [Interview – Service Quality Leader]

Facing the possibility that business as usual would lead to additional fatalities, the team was dealing with one of the most extreme types of problem encountered in any environment: the loss of human life. The severity of the situation was immediately recognised by the government authorities, which put even more pressure on the team and created an even greater sense of urgency:

> So, there was a huge focus from the authorities, and they sort of came back to us and [said that our] rescue preparedness from the turbines is not good enough. And they could as such stop us from visiting our turbines until we have a proper rescue set up. Which means that we could not service our turbines. We could not go and install our turbines. So, they could basically stop our business. No more installation or service of our turbines. [Interview – Service Quality Leader]

There were many headlines in the press. A formal 8D technical problem-solving investigation was immediately started to understand the root cause of the failure and generate improvement actions. However, as the Quality head notes, obstacles existed and the 8D was not effective for the problem-solving:

> And there was our management team that says 'hey hold on hold on'. This might cost us millions. You are being demanded to come up with information really, really rapidly. All information …for your stakeholders, especially the authorities, but also the customers and also information that can satisfy yourself internally…. And then we can see that some of the methods that we sort of use for getting our right information, that is we run an 8D or we look at our CAPA team, [to use] those methods are not really sufficient. They are OK as such, but …. that's not sufficient in a world that is highly political… [I]f you say something to the authorities or to our customers even at the very beginning just to calm them down they will take that and they will keep that information and say 'you promised us…. you said….'. The consequences of what you are saying …. you really do not know the outcome. [Interview – Service Quality Leader]

When asked how the technical problem-solving methodology was applicable to solving this type of highly visible problem, the Quality Head pointed out how the time-sensitive nature of the problem and the lack of critical information available made the Six Sigma and 8D approaches unsuitable:

> [To implement] Six Sigma you need a lot of data and often we do not have a lot of data … so you cannot use Six Sigma for that. You need to find the root cause of it, but it's also a slow process. So there is a slow root cause identifying process and then a need for information speed. You can say those two, there's something missing. There's something that is lacking with the 8D. And then … as more and more departments get involved, you can have multiple 8Ds! Engineering, EHS, Service … So, you have multiple investigations ongoing, and you could say going on at

the same time but none of them is able to control this information environment. And with the visibility of the problem—no one works together, no one. ... it turns into fighting ... It's not just the customers, the authorities, the competitors in the business, it's also internally. You have multiple people [who] want to have information and answers and maybe they do not even want to have the same answer because they have different interests. 'It's not me!' So, the engineering would like to say we have no problem and all our turbines are compliant. Our project and service want a better rescue where you can see more anchor points, more rescue equipment to protect their people and also to give their customers confidence in what we have or what the customers have already is sufficient. So, there's different types of information that these departments are actually looking for.... and they don't get them out of the 8D. At least not before the 8D is completely finished, and then we might be able to derive some information out of the 8D. But then it's often way too late. [These severe events] strike... a business like ours and it's... like a lightning strike and the big problem, it's the mud that's flushed into our organisation slowing everything down when we need to speed up.
[Interview – Service Quality Leader]

The Effect of Human Behaviour on the Problem-Solving Process

The Quality leader had this to say about working on a technical engineering problem-solving team with a problem of this magnitude and what is lacking in the current methods. He identified several challenges including information sharing and communication, and the speed of the problem-solving process.

> When you are setting up the team you really should not just look at the technical skills of persons. Well of course you need to have some good technical skills, but you also need to have someone with, say, a political flair.... We definitely need a politician that can go in and say a lot without saying anything at all, if you know what I mean. Just to buy us some time to give us room so we can figure out what exactly has happened because as I said earlier if we go out and say 'look we have made a design mistake' then, these customers, the authorities want us to check all other turbine designs to make sure that they are OK, and they force us to update the turbines. They force us to repair all the turbines, and it's quite costly and it might not be needed as the root cause is not known.
>
> So, we should be really aware of what we are saying at the very beginning if we say all our turbines are OK. And it turns out that they are not OK, then were we in good faith when we said that or did we know that we had a problem? That means that we were in bad faith that could be unsafe [and] authorities find[s] out that we actually knew that the turbines were unsafe. But we just told others that they were safe. We could also go to jail and not just get fired. So, we need to be careful what we are saying in the beginning. We need a communicator expert, not just a Black Belt [Six Sigma expert]. With this problem, so, in this case it actually turns out that, yes, we had made a mistake, you know, in our engineering department.
>
> The problem is we do not have anything to control the political game around this. The information flow that sticks together with this [is about] who is dealing with what and who is sort of giving the information. And that is missing and if the 8D or the six sigma or the CAPA or the whatever we are using is not sufficient

> ... that's [the technical problem] only maybe half the part of a problem.
>
> And then it's often gonna be the head of warranty is saying something, the head of technology is saying something, the head of service is saying something.... And of course, they speak from out of their own point of view and sometimes say. Most of the time we could face that they actually are not saying the same thing. But again, these tools, Six Sigma and 8D, do not help us in regard to that. That's because they would be fighting internally and their boss #1 wants to have one outcome and boss #3 wants to have another outcome. One problem should be one responsibility. [Interview – Service Quality Leader]

The Service Quality Leader, speaking on the mismatch between traditional technical problem-solving approaches and severe problems with a time-sensitive element:

> 8Ds are not fast. It's not a fast model. It's different steps that you need to go through and then you work your way to the end but it's not enough in a high-speed environment. It's too slow you could say. It cannot bring you the answers and information. [Y]ou cannot just put a hundred people on an 8D. I mean that doesn't make it faster. It's simply the process that you need to do. The information that you need to gain and the way that you need to manage the information. [Interview – Service Quality Leader]

The Service Quality Leader explained how during these types of extreme events, the stakes are high and the consequences potentially life altering. He described how the problem-solving process often becomes an exercise to assign blame because the traditional approaches offer no protection for employees who take ownership or responsibility for the problem:

> If the root cause points [to] you, you think you need to pay a billion euro and it scares the shit out of you, that you'll lose your job or go to prison. You try to fight and say 'Ah that can't be. It's not my entire fault. ... [A]nd I'll be sure that this is the right root cause. Why haven't you investigated this or why have you not made these calculations? Are you sure that is not this?' So when the final decision is to be made of who is to pay because that's more or less always what we talk about, who is to pay for this mistake, then no one puts up their hands and says, 'I'll pay for it'. No one does, ever! It's hide and seek and a nest of lies.
>
> I've often seen that when we [the problem-solving team] are running these projects and we want to do it our best .... actually trying to get the root cause, there's no one to shield them. ...We need some procedures or ways of shielding these people working in these projects from all the pressures that are involved from the various functions. [Interview – Service Quality Leader]

The leader of the product integrity team responsible for problem-solving these types of severe events described the experience:

> Unfortunately, I have been there heading the technical investigation of a couple of events with fatalities ... these are examples of situations where the political agenda might conflict with the technical truth. The technical part is actually less complex than the political part.
>
> In this type of an investigation in a multinational company there are so many functional lines and disciplinary lines and that becomes a very blurry picture, who actually should do what and who actually failed in a situation like that.... I think it would be too complex to firmly investigate. It became so political that I could immediately feel that my support in the organisation disappear[ed], and I was on my own. So I asked to be removed from the investigation or I would have quit the company. I had to

step out of this investigation because I could simply not cope with the political agenda in this case where basically some people were just trying to cover their ass. No one cared about the root cause, only themselves. Everyone ran scared. And I can simply not participate in that when…. humans [are] getting hurt to that level. [Interview – Former Product Integrity Leader]

Case Study Highlights

The problem described here encompasses the following characteristics:

1. The problem was triggered by a severe technical event: a technical failure led to a human fatality.

2. The problem was highly visible in the organisation: the incident was covered in the media and ongoing criminal investigations by local law enforcement, federal investigation from workplace safety agencies (e.g., OSHA in the U.S. / HSE in the U.K.), civil investigations launched by relatives of the deceased, and the ever-present threat to company leaders of corporate manslaughter action were all in play.

3. Information needed to determine the root cause was deficient: this was a one-off event with limited physical evidence for analysis.

4. The problem was time-sensitive: the local government authorities threatened to prevent technicians from servicing the turbines until the root cause was known and the turbines were proven safe.

5. Leaders used their influence and power to derail the root cause analysis, and place blame on the team.

In this scenario, 8D was started, but the process was influenced by people using political power to shift blame. There is agreement among all three interviewees that human behaviours in a high-pressure situation negatively impacted the investigation. The process was controlled by senior leaders acting in their self-interest, which delayed a resolution to the problem, weakened the problem-solving process, and had psychological and emotional impacts on members performing the problem-solving. The team did not have the tools it needed to account for, mitigate, or respond to these types of human behaviours.

Chapter conclusion and next steps

Part 1 of this book identified the characteristics of a category of problems that inhibit the efficacy of traditional problem-solving methods in the context of a severe event (S.H.I.T.) which give rise to behavioural responses that traditional technical problem-solving approaches are not equipped to address. These social mechanisms of Pressure, People, and Politics hijack and sabotage the problem-solving process when S.H.I.T. conditions are present. The identification of a S.H.I.T. problem is the first step that will empower companies to navigate and solve severe technical problems faster and more effectively, and to avoid costs, both economic and human. Traditional problem-solving does not stand up to today's S.H.I.T. conditions. It's time for a new approach. The next chapter will provide a solution to S.H.I.T. problems.

PART 2

Goal: Develop and deploy a response to S.H.I.T. problems that will reach successful resolution and promote continuous learning

CHAPTER 4 The Solution: T.R.U.S.T.

In the case of a severe event that is highly visible, where information is deficient and a resolution is time sensitive, traditional technical problem-solving approaches are undermined by human psychological and behavioural responses (3P). To mitigate the effects of these human responses, we need a solution that incites the opposite response. Part 2 of this book draws on a voluminous and extensive body of academic literature in the fields of Sensemaking and Psychological Safety. This research reaches back decades and covers a wide range of topics. For this book, I've identified the findings that are particularly useful in formulating a response to S.H.I.T. problems. Many of these ideas come from two luminaries in these fields: Dr. Karl Weick and Dr. Amy Edmondson. Dr. Weick is an American organisational theorist and professor at the Ross School of Business who is credited with introducing the concept of sensemaking into organisational studies. If Karl Weick is the forefather of Sensemaking, Amy Edmondson is the cynosure of Psychological Safety. Dr. Edmondson is professor of leadership at Harvard Business School and is a prolific writer, speaker, and thought leader on the topic of creating Psychological Safety in teams.[36] References are included in the discussion of these topics, referring readers to many of the specific publications included in the bodies of work produced by Professors Edmondson and Weick, but they both continue to innovate and share new ideas. I encourage readers to engage with their current and latest works.

---

[36] https://amycedmondson.com/

Understanding human response to severe events through the lens of Sensemaking

The brain is an active inference system. The academic field of 'sensemaking' provides a framework for understanding the human response to unforeseen, unanticipated, or unexpected stimuli. Sensemaking is the ongoing action of making sense of a situation until the situation is understood appropriately in order to make possible sensible choices and action. People engage in sensemaking when they encounter implausible events. For example, a worker has followed standard procedure, has followed all the rules, but the product is bad, and s/he must make sense of something that s/he can't understand in order to figure out what to do next.

Once a severe event is triggered, in an environment full of ambiguity and complexity, the scope of the problem itself becomes unclear. The difficulty of problem-solving and learning in a crisis is compounded by the event being unexpected and with large impact: the world has not performed the way it should, and individuals and teams lose their perspective and ability to respond. As Weick puts it, 'sensemaking is the process of organising flux'.[37] Weick borrows from classic Greek philosophy the term 'cosmology episode' to illustrate the impact a severe event has on people: 'A cosmology episode feels like "vu jàdé", the opposite of déjà vu: "I've never been here before, I have no idea where I am, and I have no idea who can help me"'.[38]

---

[37] Weick 2009: 134
[38] Weick 1993

Sensemaking is not 'making decisions', it is the foundation needed for effective action.[39] Sensemaking is a lens to look at severe events and how human response is intrinsically impacted under these conditions.

Prior academic research in sensemaking has empirically studied similar phenomena in volatile, high hazard, uncertain, and complex environments to induct theory on why failures occur and how firms can minimise the impact and increase the firm's chances of survival. For example, Karl

---

[39] Hodgkinson and Starbuck 2009

Weick's cumulative body of work has applied a sensemaking lens to severe events such as the Tenerife air disaster, the Mann Gulch disaster, and the Bhopal disaster.[40] This established literature offers insights into the causes and process of understanding, reacting to, and managing severe and unexpected events. More recently, other scholars have used and applied sensemaking to understand and explain some of our world's greatest severe events: the COVID-19 pandemic,[41] the failure of sensemaking during the Flint water crisis in Michigan,[42] the Australia Black Saturday all-time worst bushfire disaster,[43] the Global Financial System meltdown,[44] the Yarnell Hill Fire, one of the U.S. fire services greatest fatal events,[45] and the crash of Air France Flight 447 killing the crew and all passengers.[46]

Following the logic of how sensemaking develops and can be controlled during a severe event, this section focuses on the elements critical for successful sensemaking. The elements of leadership, communication, expertise, and enactment can mitigate the loss of order that accompanies severe events.

Leadership: The leader plays a central role in influencing team behaviours/responses to a severe event, giving sense to the team; framing

---

[40] See Weick 1990; Weick 1993; Weick 2010
[41] Christianson and Barton 2021; Sahay and Dwyer 2021; Galbin 2021; Stephens et al. 2020
[42] Nowling and Seeger 2020
[43] Dwyer et al. 2020
[44] Hollerer et al. 2018
[45] Williams and Ishak 2018
[46] Berthod and Muller-Seitz 2018

and labelling the problem; and empowering team members to find solutions.

Communication: The leader uses language and storytelling to foster a cycle of real-time continual communication, team action, learning from cues, and repeat aligning the team during the severe event.

Cross-functional experts: A team of experts is needed to problem-solve the event and they take actions to generate cues while avoiding deleterious impacts such as information overload and the impact of stress.

Enactment: Abduction logic is needed as the team works in a complex environment working at the edge of codified knowledge, pushing forward through adversity as an upward trajectory determines success.

Figure 13 summarises the flow of these elements for sensemaking during the event.

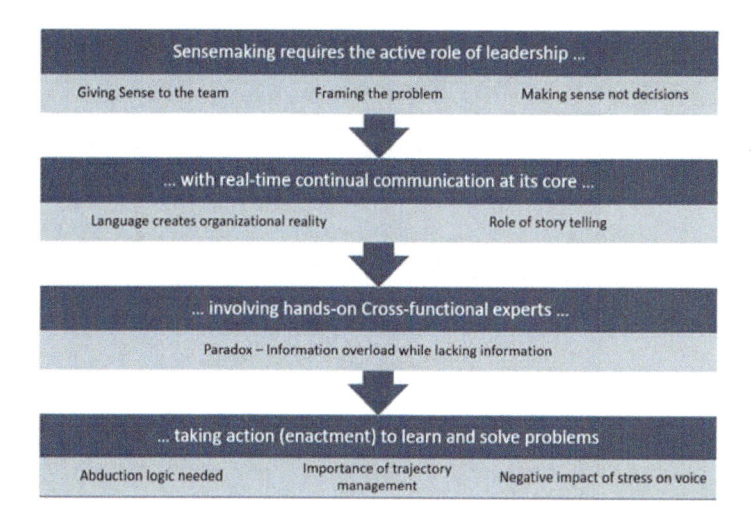

Figure 13 Sensemaking during a severe event

Each of these elements is discussed further.

Leaders

The sensemaking perspective advances the theory that senior leadership has an active role in helping people and organisations navigate severe events and can operate and manage within this context 'by socially constructing their environments, selectively perceiving some cues and ignoring others, and then enacting what gets labelled as a surprise or a crisis'.[47] Sensemaking is fundamentally a social process where leaders enact sense using their abduction skills to generate meaning from fragmented and often conflicting evidence and cues to generate plausible stories to get the team moving and responding (Weick 2010.) It is through this 'sensegiving' that leaders can lead through and beyond, in a better and

---

[47] Beck and Plowman 2009

stronger state then before the problem occurred. Drawing from the literature review of the sensemaking perspective, problem-solving during a crisis can be enhanced by active leadership to:

1) give sense to the team
2) provide active construction and frame the problem
3) empower the team to find solutions by making sense and not decisions.

Each of these leadership roles and how they will enhance problem-solving during a severe event are discussed.

Giving sense to the team

The sensemaking literature offers principles on how cross-functional empowered teams, structured with appropriate expertise and with supporting leadership and good communication, are necessary to appropriately learn through and from the crisis.[48] The unit of meaning in sensemaking is the relations between cues and the frame.[49] Leadership in sensemaking literature is an active hands-on role, using the team as the leaders' hands to generate actions and knowledge. Referred to in the sensemaking literature as sensegiving,[50] these actions are interpreted and filtered by the leader, creating the narrative for the team to make sense of its situation and move forward. 'Action unattached to a narrative is

---

[48] Christianson et al. 2009
[49] Weick 1995
[50] Gioia and Chittipeddi 1991

senseless'.[51] Leaders interpret the severe event, and then influence how others interpret those same events.

Organisational psychology theorises that people and organisations possess limited capability for dealing with 'diverse stimuli'.[52] Humans cope with these intrinsic limitations of their abilities by focusing awareness and attention on a limited set of topics and problems. This focus is determined and propelled by whatever is deemed salient for the greater organisation. In sensemaking literature, severe events are enacted through their salience and prominent impact.[53] Thus, the salient impact of the severe event activates the sonar of senior leadership as the alarm of the potential for calamity is raised through this 'enacted salience'. Sensegiving thus allows a leader to create meaning for those impacted and, as a first response, a leader typically asserts her narrative interpretation on what has occurred, with the intention of influencing how others interpret those events.[54] Sensegiving can be summarised in three words: labelling imposes order. Accordingly, the leader is attempting to give sense to the situation, and in doing so, is guiding others toward 'a preferred redefinition of organisational reality'.[55] Once the problem has been categorised and defined by the leader (i.e., labelled), power is then transferred to teams assembled with the proper expertise and empowered to diagnose cause and develop further action (the basis for problem-solving). Again, sensegiving can be summarised in three words: labelling imposes order.

---

[51] Weick 2009: 202
[52] Weick and Sutcliffe 2006
[53] Weick and Sutcliffe 2006
[54] Lampel et al. 2009
[55] Gioia and Chittipeddi 1991: 442

Sensegiving is a proactive action occurring during the early stages of the crisis.[56] Leadership is necessary to responding since when the severe event occurs, the team is immobilised: 'the orderliness of the universe is called into question because both understanding and procedures for sensemaking collapse together—people stop thinking and panic'.[57] In these conditions, the leader 'embodies the possibility of escape'.[58] Assuming the role of a sensegiver, the leader imparts faith and trust in the team to get them moving and acting. Drawing on the confidence of the leader, problem-solving starts as the team regains composure by monitoring cues from the environment, which triggers learning what happened and formulating working hypotheses on what must come next. Sensegiving is about providing the answers for two fundamental questions: 'what's the story here?' and 'now what should we do?'[59] Labelling imposes order; framing narrows the focus on where the team must act.

Framing the problem

If the event is regarded as a crisis, a sense of alarm is generated; if the event is instead labelled as a 'continuous learning' opportunity, patience follows.[60] Multiple industries have already adopted neutral labelling of a gap in expectations or an unknown as common practice. In medical operations the word 'error' is used to reflect hapchance as inevitable as human nature; in engineering operations with the ISO 9000 Quality Management System reference, the term 'nonconformance' is used as a

---

[56] Chia 2000
[57] Weick 1993
[58] Weick 1995
[59] Weick et al. 2005
[60] Thompson 2014

neutral unbiased attempt to reflect a deviation from a standard. 'That is the whole point of sensemaking. Once something is labelled a problem, that is when the problem starts'.[61] And once the problem starts, so begins the problem-solving process.

The leader defines what is and what is not critical to address in relation to the severe event, as well as the boundaries of the severe event problem the team must solve, through a process of naming the things to focus on and framing the context for that focus.[62]

Sensegiving is about branding the event, categorising, and naming the problem which 'stabilises the streaming of experience'.[63] Framing is to codify and construct the problem statement explicitly from the environmental context. This creates the frames within which cues are noticed, extracted, and used for further actions.[64] The framing of the problem also reveals which experts are necessary to the problem-solving team as the leader seeks competence in highly technical issues.

Sensemaking is induced by an incapacity to extrapolate current actions and to foresee their consequences due to the uncertainty and ambiguity around the event. Lack of framing will increase the likelihood the crisis will spiral out of control as the team's uncertainty and ambiguity will lead to a loss of focus. When framing, the leader applies boundaries to the problem, moving the team together forward and faster for action.

---

[61] Weick 1984: 48
[62] Schön 1983b in Weick 1995
[63] Weick et al. 2005
[64] Snow et al. 1986

To summarise, the shock of the severe event causes people to lack understanding on what happened, 'what's the story'. The enacted salience draws management attention. Once triggered, predictable sensemaking processes and steps are activated and can be influenced[65] and manipulated. The problem is labelled as a priority and framed as a challenge the organisation can and will overcome, signalling the need to understand and learn to go forward. The leader supports this effort by obtaining information and interpreting signals from the environment, using these cues to form working hypotheses, providing the foundation to restore order and 'make sense' of the event and continue further actions, all while delivering perpetual real-time communication to the teams.[66]

Making sense (not decisions)

Leaders following the principles of sensemaking do not make decisions, they make sense by following cues and empowering the expert team for actions. While framing the event as a matter of meaning, as opposed to making decisions, the emphasis is shifted away from the individual decision maker towards action and learning.[67] The following quote from Paul Gleason, an American firefighter, highlights the salient difference between decision-making and sensemaking:

> If I make a decision, it is a possession; I take pride in it; I tend to defend it and not to listen to those who question it. If I make sense, then this is more dynamic, and I listen and I can change it. A

---

[65] Sandberg and Tsoukas 2015
[66] Maitlis and Christianson 2014
[67] Snook 2000

decision is something you polish. Sensemaking is a direction for the next period.[68]

Decision making during a severe event consumes the time and effort of justifying the decision to others, which may lead to escalating commitment when the decision is challenged. In a severe event, focus must be on reacting to the event by understanding cues and formulating a response. Context is important as the cues provide an impetus for actions: moving to understanding root cause, formulating the problem statement, and mobilising the team. As the crisis leadership expert Charles Perrow writes, 'to understand ... we have to examine the context of the failure.'[69]

Following the formula that a severe event begins the sensemaking process with three elements, 'a frame, a cue, a connection', the United States Forest Service firefighters have developed a model to support leaders with making sense, not decisions, during a severe wildland forest fire. Referred to as STICC (Situation, Task, Intent, Concern, Calibrate), the model has been adopted in fields such as emergency medicine, and by the U.S. military, where time pressures and errors can lead to severe consequences.

---

[68] Paul Gleason, cited in Weick 2009, *Making Sense of the Organization,* 75
[69] Perrow 2004: 10

Wildland firefighting Leadership Sensegiving – STICC – situation, task, intent, concern, and calibrate				
o Situation – 'Here's what I think we face.' The leader summarizes the current problem or issue	o Task – 'This is what I think we should do.' The leader explains their plan for addressing the problem.	o Intent – 'Here's why.' The leader provides the rationale behind their plan.	o Concern – 'Here's what we should keep our eye on.' The leader identifies potential issues or problems that could arise in the future.	o Calibrate – 'Now talk to me.' The leader invites feedback or questions

2011 – Marlys K Christianson, Kathleen M Sutcliffe - Becoming a high reliability organization

Figure 14   STICC method for sensemaking in a crisis situation

To summarise, the leader applies framing to make sense and define the scope of the problem to be addressed and empowers (not micromanages) the team by asking questions during the severe event, prompting actions. Even when leaders adopt these approaches, the literature discloses certain pejorative difficulties that can derail and prevent successful sense making.

Threats to successful leadership during a severe event

Sensemaking literature identifies a variety of threats or pitfalls that proactive leaders should be wary of in times of crisis; threats that can limit or prevent making sense of the situation. For example, the combined pressures of uncertainty and ambiguity can lead managers to snatch decisions from lower-level employees in a bid to exert their control.[70] This retrenchment of authority from experts to management leads to less action and more confusion for the team.[71] Sensemaking researcher William Starbuck, in studying leaders' cognitive reactions to severe events (crises), inducted the following:[72]

> The reactions to uncertainty include wishful thinking, substituting prior beliefs for analysis, biasing probability distributions toward

---

[70] Roberts et al. 1994
[71] Weick 1998
[72] Starbuck 2009: 930

certainties, searching for more data, acting cautiously, and playing to audiences.

Finally, the literature describes how a lack of leadership can hinder sensemaking and the ability to respond and act in a severe event crisis. Tension builds as the team is left on its own without a clear sense of how to act in response to the severe event, leading to 'regression to more habituated ways of responding, the breakdown of coordinated action, and misunderstandings in speech-exchange systems.'[73]

Further, allowing pressure to build may lead the team to become focused on self-preservation and not on responding to the environmental cues that assist in determining the root cause of the crisis. For example, in a study of medical emergency teams placed in a simulated unplanned crisis event (the simulation had a piece of medical equipment failing during a medical procedure leading to patient fatality) sensemaking researcher Marlys Christianson notes the condemning effect of pressure on the team's performance to solve the problem. Increased stress during a severe event results in teams having a narrowed focus, the inability to notice other cues, and an over-reliance on early hypotheses due to placing focus on confirming evidence.[74]

Leaders have the important role of giving sense to the team by framing the problem and focusing on making sense, not top-down decisions. As is now explored, real-time continual communication is at the core of this type of leadership.

---

[73] Weick 1990
[74] Christianson 2019

Communication

Communication is the glue for combining leadership intent with team, trust, and action. A severe event is accompanied by a dearth of meaning and deep instability. Individuals and leaders need to make sense of what is happening and often intuitively construct a narrative to satisfy the need for the organisation's understanding. 'My act of speaking starts the sensemaking process'.[75] Sensemaking involves events turned into a state that is explicated in words and dialect to serve as a 'springboard to action'.[76] Using storytelling to illustrate the event and incorporating language the team will understand is an essential tool for managing the unexpected.[77] After the shock of the severe event crisis occurs, the team is immobilised and unsure of what to do next.[78] Communication is the first tool[79] in the leader's toolbox to give sense to the team and spring forward. The world will not end, and additional actions must be launched: analyse the situation and apply corrective manoeuvres.[80] Using frames to guide conduct by facilitating the interpretation of cues turned up by that conduct, a hypothesis is formed and enables the root cause analysis to begin. Physical action is the further trigger for successive conversations and language that interpret the severe crisis event.[81] Sensemaking is communication; communication brackets the context by placing discrete labels on our continuous flow of experience.

---

[75] Weick, 1995: 12
[76] Weick and Sutcliffe 2005
[77] Weick and Sutcliffe 2015
[78] Sandberg and Tsoukas 2015
[79] Lampel 2009
[80] Beck 2009
[81] Weick 2009

For the practitioner, a lesson from sensemaking is the importance of actual meaning construction during the event. In other words, an event is not a severe event until defined as a severe event:

> Constructing sentences to express statements about experience imposes discrete definitions on a subject matter that is continuous. One cannot report in a sentence an observation about experience without a concept that structures what one is observing. Observation statements describe not perceptions but planned perceptions. Data are not given by experience, but by the concept of the language used to interpret it. Observational language imposes discrete boundaries on the continuity of the phenomenal world so as to define concrete, individual events in that world.[82]

Sensemaking provides strategies such as how to understand the situational context, utilise the tools of communication, and turn the event into an opportunity for learning, improving organisation routines, and becoming a more resilient organisation.[83] Language provides sense through framing and action.

Language creates organisation reality

'Organisations are talked into existence.'[84] Groups obtain their substance and legitimacy from the content of the discussions that are generated and take hold. This firms up when leaders frame the topics, the conversations are repeated, and the content becomes the status quo. Communicating the boundaries of the crisis makes the abstract real. It determines what cues are chosen as relevant, and the scope of the crisis to be addressed. 'These

---

[82] Weick 1995: 107
[83] Weick and Sutcliffe 2015
[84] Weick 2009

conversations create organisational reality; they don't just represent an entity that is already there.'[85] Meanings materialise and ironically the crisis is invented in a sense through language and labelling the cues from the severe event. It is this same labelling that promotes sense as we create theories from the stream of endless perceptions, and our beliefs are codified into words and descriptions that we can meaningfully communicate and share with others:

> Sensemaking is a social process that edits, abridges, simplifies, and brackets our first-hand experience with flux. And the social character of sensemaking determines how well the unexpected can be managed.[86]

Severe events come like a lightning strike where people scramble for action. Sensemaking goes past interpretation and necessitates the active authoring of events and socially constructing structures for understanding, as individuals play a part in authoring the very situation they struggle to understand.[87] The leader must unify the cues into a narrative the audience will understand; the leader needs to tell the story.

Storytelling

> If accuracy is nice but not necessary in sensemaking, then what is necessary? The answer is, something that preserves plausibility and coherence, something that is reasonable and memorable, something that embodies past experience and expectations, something that resonates with other people, something that can be constructed retrospectively but also can be used prospectively,

---

[85] Weick and Sutcliffe 2015: 36
[86] Weick and Sutcliffe 2015: 33
[87] Maitlis and Christianson 2014

something that captures both feeling and thought, something that allows for embellishment to fit current oddities, something that is fun to construct. In short, what is necessary in sensemaking is a good story.[88]

As discussed, in the sensemaking literature, framing is used at the onset of a severe event to recover and lessen the impact by framing the event as an opportunity for learning and improvement. A corresponding and necessary sensegiving tool for the leader to utilise for the duration of the severe event is storytelling. Stories are a framing device for meaning construction, thereby turning 'the unexpected expectable, hence manageable'.[89] Stories serve as 'guides to conduct', the frames created by leaders that facilitate the understanding of cues, helping to guide behaviour during the severe event.

The literature discusses the value of storytelling for the organisation experiencing the severe event, emphasising that accuracy is not the highest goal. Rather, a story functions as a tool that can preserve plausibility and coherence, resonate with people, be used prospectively, and capture both feeling and thought. In the context of sensemaking, the elements of a good story must be reasonable and memorable, something that embodies past experience and expectations, constructed retrospectively, and something that allows for embellishment to fit current oddities.[90]

---

[88] Weick 1995: 60
[89] Weick 1995: 127
[90] Weick 1995

A story connects dissimilar components together long enough to galvanise the team to act. The story must be credible, to allow persons to make retrospective sense of the event, and sufficiently compellingly enough 'that others will contribute their own inputs in the interest of sensemaking'.[91] The leader can frame the story as one of opportunity or innovation (e.g., 'we are on the bounds of technology, and we are all in it together'), and use it to energise the team and explain next steps, all while making sense of the problem. A story can be used to frame the severe event as a challenge, as a 'team problem to solve', thereby reducing pressure and energising team members to work collaboratively. Thus, stories are a template for diagnosis, reducing provocation, moving past shock, and triggering the team to look for cues and promote understanding.

While stories are necessary as a tool for leadership to frame the problem and organise momentum, the momentum must turn to action: a cross-functional team of experts is needed to act.

Cross-functional experts

> There are limits to what novices and advanced beginners know, which means they miss or intentionally ignore more cues. Missed cues mean that subsequent action may make things worse and lead to more frequent and more intense subsequent breakdowns. ... Noticing occurs in the service of resumption, recovery, and resilience. And the size and content of the response repertoire determines noticing.[92]

---

[91] Weick 1995: 61
[92] Weick 2009: 78

Cross-functional experts are needed when dealing with a severe event. Following the logic of the sensemaking perspective just described, the leader has applied sensegiving to frame the problem and then used communication to get the team moving: extracting cues, beginning recovery efforts, and containing the severe event. Amid the uncertainty and ambiguity of the crisis, empowered cross-functional teams are needed. For example, following the Columbia shuttle disaster, an independent board investigating the response efforts advised the National Aeronautics and Space Administration (NASA) to rely on the technical knowledge of their experts, recommending that NASA:'…restore deference to technical experts, empower engineers to get resources they need, and allow safety concerns to be freely aired.'[93]

For sensemaking to be effective in a crisis, experience means more than rank. Experts can help make sense in a crisis because they react to relevant information or 'clues' faster, are able to pay more detailed attention to context in the environment and recognise the patterns as the severe event unfolds, aiding in the formation of hypotheses for further actions.[94] These abilities are particularly valuable during a crisis because speed is needed to gain trajectory and minimise the impact of the severe event:

> Men who can immediately sense the potential problem can indeed make a quick decision to alleviate the problem or effectively decouple some of the technology, reducing some of the consequences of errors … decisions migrate around these

---

[93] CAIB 2003: 203, https://www.nasa.gov/columbia/home/CAIB_Vol1.html
[94] Weick 1995

organisations in search of a person who has specific knowledge of the event.[95]

In the midst of a severe event, the sensemaking literature argues that decision-making should migrate to experts and that their expertise and experience should be more highly valued than rank.[96] Experts can be empowered when leadership overtly slackens and releases hierarchical constraints, energising those with the skills to address the problem. This symbolises the sensemaking belief that within the context of the problem, the essential capabilities lie somewhere in the organisational system and relying on that team is the right course of action. The problem-solving team should include a group of experts, a 'team of teams', representing multiple cross-functions including manufacturing, engineering, suppliers, and even customers.

The role of active leadership continues with the cross-functional team. Team members come from diverse groups, with different organisational objectives and different cultures., and it is difficult for them to operate as one team.[97] The meanings of cues can be understood differently based on functional background and training, resulting in 'enormous effort required to create cross-functional teams whose members share even a modest number of meanings'.[98] Specialists try to pull the cues unintentionally and automatically toward their field and area of expertise,[99] and this pejorative effect can be increased by the effects of stress and information overload.

---

[95] Karlene Roberts et al. 1994: 622
[96] Weick and Sutcliffe 2015: 126
[97] Schaafstal et al. 2001
[98] Weick 1995: 113
[99] Hashem 2003

The active leader can help avoid these pitfalls by unifying the team through active participation and communication.

And once the problem is solved, the root cause having been determined, and actions are put in place, the team of teams is disbanded and the world has regained composure: the problem will end. Closure can be as important as the expert knowledge itself. If you have a team of amateurs who have never undergone problems in the past, pure technical knowledge is a poor substitute for practice. 'If you experience infrequent setbacks, then you have little experience opposing and ending such events. And if you are spared from the full force of collapse, failure, and disappointment, then you never learn the lesson that bad things come to an end. If you fail to learn the lessons of closure, then bad experiences, when they eventually do occur, can seem overwhelming'.[100]

Teams that include empowered experts are best suited to act during the severe event. But even experts are human and can be susceptible to human failures.

Severe Event Paradox: learning through the severe event in the face of simultaneous information overload, uncertainty, and informational ambiguity

The sensemaking literature argues that non-routine problem-solving is needed in highly stressful situations because of information ambiguity, uncertainty, and overload.[101] In a severe event, information overload can

---

[100] Weick and Sutcliffe 2015: 118
[101] Schaafstal et al. 2001

have a deleterious impact. Information overload can transform experts into a more regressed state with a reduction in cognitive abilities and processing.[102]

When synchronisation and alignment is needed the most, due to 'the breakdown of coordination ... the group acts less like a team than like individuals acting in parallel'.[103] Severe impact and highly visible decisions coupled with execution, time, and performance pressures, when combined with information deficiencies, 'trigger a regression to a less expert stage'.[104] To visualise the pejorative effect of the strain on a cross-functional team, Weick uses the term 'advanced beginners'[105] to reflect sub-functioning experts with a reduction in the ability to process and respond to the external environment as a result of regression to a less mature state. The crisis team risks skipping over important data, and unconsciously seeking bias confirmation from preliminary hypotheses to simplify the world once more. It is under these conditions that the leader must again actively protect the team, provide additional sensegiving through framing the problem as an opportunity, and use storytelling for the larger organisation as necessary.

As mentioned earlier, especially during these moments, decisions should not be made that will further affect confirmation bias and 'polish the decision', instead the leader must continue to make sense and keep the team flexibly responding to cues. Traditional problem-solving and project

---

[102] Barthol 1959
[103] Weick 1990
[104] Barthol and Ku 1959
[105] Weick 2009

management promotes learning procedures at the end of the project or project phase. Instead, sensemaking treats the severe event as a 'brutal audit' of the firm's management system, and learning is applied through the event, thereby mitigating the bias of the benefit from hindsight. 'The learning is less that of "lessons learned" away from action and more of "skills learned" during the action. Old routines, which can or had been vague, now become clearer as they are turned inside out.'[106] By learning through the event, leadership signals to the organisation that the goal is not to place blame, but rather to learn from the event: mistakes happen and this is an opportunity to apply continuous improvement.[107] This is important as the team managing the severe event should conduct the investigation in an open manner and not be blocked by the paradox of too little and too much information. By acting and remaining receptive to the environment the cross-functional team learns and continues to problem-solve the crisis.

Enactment: creating knowledge through action

As has been discussed, specific struggles at sensemaking likely occur 'when the current state of the world is perceived to be different from the expected state of the world'.[108] Our analysis in chapter 2 supports this perspective: for the individuals responding to a severe event, there is a perception of no discernible approach to engage and connect with the world and others around them. In these events there is an existential experience change; the world has become unintelligible. 'Enactment is

---

[106] Christianson, Farkas, Sutcliffe, and Weick 2009
[107] Weick, Sutcliffe, Obstfeld 2005
[108] Weick 2009: 131

premised on the idea that people play a key role in creating the environment in which they find themselves.'[109] When the crisis is triggered by a severe event and all sense is lost, the leader gives sense by using language to frame the problem as an opportunity for improvement and uses storytelling to communicate the boundaries of the problem. A cross-functional team is assembled to pay attention to cues and generate hypothesis on root cause. The final concept of sensemaking to be discussed is the concept of enactment: the environment is produced through action. If the opening query of sensemaking is 'what's the story here?', the second, correspondingly vital question is 'now what should we do?'[110]

The word 'enactment' was originally defined for the sensemaking construct because enactment suggests there are close parallels between what leaders do during a crisis and what legislators and lawmakers do. Both groups construct reality through acts. When people enact laws, they take underdefined space, time, and apply action to draw lines, establish categories, and coin labels that create new features of the environment that did not exist before. Sensemaking 'involves enacting more or less order into ongoing circumstances'.[111]

In contrast to the linear logical approach of traditional technical problem-solving, enactment means:

> Action and talk are treated as cycles rather than as a linear sequence. ... Because acting is an indistinguishable part of the

---
[109] Maitlis and Christianson 2014
[110] Weick and Sutcliffe 2005
[111] Weick et al. 2005: 409

swarm of flux until talk brackets it and gives it some meaning, action is not inherently any more significant than talk, but it factors centrally into any understanding of sensemaking.[112]

Enactment has been defined to illustrate the idea that in a crisis, when the elements surrounding the event don't make sense, the actions people take are socially constructing and 'bring events and structures into existence and set them in motion'.[113] This is a salient feature of the sensemaking perspective. 'Sensemaking "starts" with knowledge by acquaintance that is acquired through active exploration. Active exploration involves bottom-up, stimulus-driven, online cognitive processing through action.'[114] Enactment is not based on inductive or deductive reasoning where the actors passively sit back analysing data, forming abstract experiments to test hypotheses under static conditions. Rather, abductive logic is needed in a crisis with equivalent options with no clear probability.

Abductive logic is needed in a severe event

Using traditional problem-solving methods like 8D, based on statistical process control, knowledge is generated from careful application of the scientific method. Hypotheses are formed on the relationship between variables, experiments are conducted, samples are produced to measure the process influence, and conclusions are drawn from the samples to the overall population parameters. Statistical laws like the central limit theorem govern this inductive logic process, including the minimum amount of trials needed (normally more than 30) and how many replicates

---

[112] Weick 2009: 136
[113] Weick 1998
[114] Weick 2010

of the experiment are necessary to claim scientific power and probability significance.[115]

In the sensemaking literature, knowledge is generated through action. Rather than using statistical induction, in a severe event action is enacted on the environment using 'abduction to generate plausible meaning from fragmented evidence'.[116] Abductive logic is used to form and evaluate hypotheses to make sense of bewildering and perplexing facts and to find the most likely explanation.

With abductive logic 'good' is not the enemy of 'great'; action generates cues to learn for further action on the route to root cause. As the cross-functional team of experts continues to test its working hypothesis from direct knowledge, the experts learn one failure at a time. Each time moving closer to the root cause of the severe event.

Trajectory management

In the sensemaking literature actions matter, and first actions matter most. Trajectory management suggests that if teams can recover from the initial shock and work together quickly, that begins a virtuous cycle: 'Initial responses do more than set the tone; they determine the trajectory of the crisis.'[117] Trajectory management continues as the repeated ongoing process throughout the problem of revising conditional sense-making to incorporate new cues and new learning.[118]

---

[115] Durivage 2016
[116] Weick 2010
[117] Weick 1998
[118] Christianson 2019

Trajectory management through enactment stimulates 'decision making to local frontline experts who are empowered to solve problems'.[119] Action is important as time is critical in a severe event and should not be wasted. One can imagine the potential opportunity cost incurred if traditional problem-solving were followed, with its continual reviews and explicit governance requirements, since time lost can have negative impacts and determine the success or failure in a crisis. As the crisis develops, action and cognition fold together as the teams listen, learn, and react to the environmental cues. Benefits rebound as the severe event is contained, and resources spent on the problem-solving event are minimised. As a key lesson from enactment, understanding is facilitated by action and 'reluctance to act could be associated with less understanding and more errors'.[120] This reveals that the value of sensemaking differs not only on how groups remake sense but also on how they connect as a team: 'more-effective teams monitor and rapidly interpret cues, confirming them with others and evaluating changes over time.'[121]

The stress of a severe problem hinders cognitive capacity and deters communication

Stress is the adversary of action, but stress—and its many negative impacts—is a natural part of severe events.[122] Stress hinders response to the severe event as cognitive capacity is reduced and distraction increases:

---

[119] Christianson and Sutcliffe 2011
[120] Weick 1998
[121] Christianson 2019
[122] Schaafstal et al. 2001

cues are missed, reducing acting and sensemaking.[123] This can be a cycle that can derail the sensemaking effort.

This impact is further exacerbated during a crisis moment: high stress leads to lack of communication and sharing of information.[124] This finding was confirmed in a study of medical errors in an emergency medical setting.[125] The author described how communication failures were contributary factors in 91% of errors. The communication failures arose from vertical hierarchical differences between roles, interpersonal power, and conflict.

A crucial understanding from enactment is the lesson that during a crisis, teams must have speech and language: 'A key to learning from severe and unusual events is to experience them richly by encouraging multiple voices, perspectives, and interpretations.'[126] While this statement is certainly true and valid, the sensemaking perspective is complemented by applying the concept of psychological safety.

Psychological safety offers prescriptions and mechanisms to promote many of the characteristics needed for effective sensemaking: empowering and giving voice to team members, improving communication, encouraging new ideas, and reducing the effect of power distance between roles to encourage a well-functioning cross-functional team of experts during a severe event.

---

[123] Weick 1995
[124] Blatt et al. 2006
[125] Sutcliffe 2008
[126] Beck and Plowman 2009

Improving human response with Psychological Safety

The theories of sensemaking delved into the factors affecting how a leader can influence individuals to make sense and act during a severe event, but there is also literature covering team and organisational factors. To mitigate the pernicious effect severe events can have on the efficacy of traditional problem-solving, the team should be made to feel psychologically safe. Psychological Safety is the belief and trust that it is safe to provide voice and engage in risk taking activities without fear. Early research into the group and organisational factors on problem-solving found that management's leadership behaviours related to how mistakes are handled, i.e., 'create an ongoing, continually reinforced climate of openness or of fear'.[127]

The behaviours of management were established to be a contributing factor into whether problems were actively hidden or ignored versus communicated and addressed by applying problem-solving techniques. Further research into the success or failure of problem-solving and consequently improvement and learning resulted in the seminal paper, *Psychological Safety and Learning Behavior in Work Teams*.[128]

Psychological safety is grounded in the belief that engaging in problem-solving can be perceived as a risky activity in the workplace. Fears of personally being blamed for mistakes, being viewed as incompetent by colleagues, and being punished result in workers not actively admitting errors, which limits organisational improvement as problems go unsolved

---

[127] Edmondson 1996: 80
[128] Edmondson 1999

and potentially escalate out of control. By performing analysis on teams objectively successful or not with continuous improvement, the concept of psychological safety was constructed:

> Team psychological safety is defined as a shared belief that the team is safe space for interpersonal risk taking. ... The term is meant to suggest neither a careless sense of permissiveness, nor an unrelentingly positive affect but, rather, a sense of confidence that the team will not embarrass, reject, or punish someone for speaking up. This confidence stems from mutual respect and trust among team members.[129]

In going deeper into the considerations for unsuccessful learning from mistakes, psychological safety researchers provide a critique and challenge to traditional technical problem-solving research. This assessment states traditional problem-solving has focused on designing 'neat' closed-loop methods and techniques for root cause analysis while not fully understanding how people behave and confront problems in the workplace. Psychological safety experts assert that the focus of traditional problem-solving on identifying preferable methods has resulted in a gap in understanding about what happens when human beings confront problems in organisational contexts.[130]

As seen in the examples of severe event problems iterated in chapter 3, traditional problem-solving methods '... are normative in nature and difficult to use in the hectic front-line environment'.[131] In order to have a deeper understanding of the limits of the traditional approaches,

---

[129] Edmondson 1999: 354
[130] Tucker et al. 2002: 125
[131] Tucker and Edmondson 2002: 89

psychological safety experts have developed language by deconstructing traditional technical problem-solving approaches into two phases: single-order and double-order problem-solving.

> Analogous to the concepts of single and double loop learning identified by organisational learning researchers (Argyris and Schon 1978), research on problem solving makes a distinction between fixing problems (first-order solutions) and diagnosing and altering underlying causes to prevent recurrence (second-order solutions). First-order problem solving allows work to continue but does nothing to prevent a similar problem from occurring.[132]

Using this first- and second-order framework, the authors argue that psychological safety enables second-order problem-solving behaviour by reducing the inherent risks of improvement efforts. They claim that, highlighting a need for change may uncover mistakes and shortcomings, which can lead to consequences for the person who raises the concerns. This is particularly relevant to organisational power, where the person raising the concern is not as powerful as the person responsible for the shortcoming.[133] Dutton and Ashford argue that being associated with problems and change efforts can be damaging to an employee's reputation.[134] All of these negative consequences mean that workers will be more likely to engage in improvement efforts if they feel protected.[135]

Over the past 20 years psychological safety has involved ideas from merely defining the absence or presence of the construct to a growing

---

[132] Tucker et al. 2002: 124
[133] Edmondson 1996
[134] Dutton and Ashford 1993
[135] Edmondson 1999

literature on the 'cognitive, interpersonal, and organisational adjustments'[136] necessary for teams to perform in the toughest of environments and successfully learn from mistakes. In addition, beyond a static stable work environment, studies of temporary work teams have highlighted that perceived differences in status between colleagues from different functions can also have a detrimental impact and prevent 'pointing out potential problems and admitting mistakes'.[137] This effect was evident in the research of the interactions in problem-solving teams between blue-collar shop-floor employees and white-collar engineering colleagues.

Several mechanisms identified in the Psychological Safety literature that are pertinent to severe events are discussed below.

Inclusive Leadership: enabling voice to minimise the impact of power differences

An inclusive leader, humble enough to ask for help, is important for establishing psychology safety. Cognisant of their power, such leaders limit threatening behaviour as 'threat has the effect of reducing cognitive and behavioural flexibility and responsiveness'[138] right when teams need to perform at their best.

Psychological safety literature asserts that collaboration among peers brings essential elements to complex problem-solving, where intense coordination may be needed, claiming that traditional managerial controls

---

[136] Pisano et al. 2001
[137] Edmondson et al. 2001
[138] Edmondson 1999: 352

fall short in enabling this collaboration. Formal organisational structures divide people by speciality, focusing attention on a hierarchal system, and leadership is needed to overcome these obstacles to psychological safety.[139]

Instead, coaching is the primary tool of the inclusive leader, because 'effective coaching is likely to contribute to members' confidence in the team's ability to do its job'.[140] As there is a historical power distance between functional teams (e.g., Quality vs. Manufacturing, Manufacturing vs. Engineering, etc.) a key capability of the inclusive leader also incudes coaching: 'coaching may require also explaining why others' input is essential to effectiveness'.[141] Everyone on the team has a job and must be safe to play; functional teams may be insular and need outside leadership to work across functions effectively.

The inclusive leader is chosen for leadership abilities and not functional expertise[142]: 'Leaders have at their disposal four software tools: emphasising purpose, building psychological safety, embracing failure, and putting conflict to work.'[143]

In psychological safety, leadership is a tangible deliverable property of the management system that must be defined and actively managed as a

---

[139] Edmondson 2014
[140] Edmondson 1999: 357
[141] Edmondson 2003: 1424
[142] Edmondson et al. 2001: 2
[143] Edmondson 2012: 78

system property: 'Just as safety is a property of a system rather than solely the result of individual skill, leadership is also a system property.'[144]

Fundamental to the understanding of psychological safety is the hands-on role of leadership. A criticism of empowerment and delegation is that it can lead to abdication of responsibility and hinder problem-solving: 'The flip side of empowerment, however, is the removal of managers and other non-direct labour support from daily work activities, leaving workers on their own to resolve problems that may stem from parts of the organisation with which they have limited interaction.'[145]

The importance of executive leadership is its active participation and mentoring to ensure the culture of a fearless work environment prevails:

> To learn from failures, people need to be able to talk about them without fear of ridicule or punishment. Managers can help create an environment where workers feel safe taking the interpersonal risks that second-order problem-solving entails, thereby making this behaviour more psychologically feasible.[146]

Leadership inclusiveness is the 'words and deeds by a leader or leaders that indicate an invitation and appreciation for others' contributions'.[147] The inclusive leader is proactive and invites participation through language and actions. Studies have demonstrated that when a real or perceived power difference exists, people will not speak up due to perceived lack of psychological safety. These hierarchical communication constraints derive from status differences and are unproductive as key

---

[144] Carroll and Edmondson 2002: 54
[145] Tucker and Edmondson 2003: 64
[146] Tucker and Edmondson 2003: 67
[147] Nembhard and Edmondson 2006: 974

information related to facts, data, and other types of information are withheld. 'A useful discipline for leaders is to force moments of reflection, asking themselves and then others, "Is this the only way to see the situation? What might I be missing?"'[148]

Status based differences are not only based on the organisational hierarchy between levels of control but often exist in profession-based roles in the workplace, defined by function and education. As with the well-documented case of doctors and nurses in the medical field, the engineers who designed a product may be perceived as the 'experts' when problem-solving for a nonconformity of that product. That perception coupled with their presence could inhibit input from other functions such as quality or production, who may have first-hand experience with the problem. Leaders are needed to actively promote a learning environment for problem-solving. 'For teams that have leaders, leader behaviour has been shown consistently in field research to be an important factor in shaping the climate of the team and in motivating learning.'[149]

Problem-solving inherently involves the risks of changing the system and challenging the status quo, and the right leadership behaviour is necessary to promote any change.[150] Otherwise, there is silence, people hide information, and teams fail to work together on the problem.[151] Power differences have real effects, such as perceptions of threat to identity and concern of adverse consequences. Instead, by applying leader

---

[148] Edmondson 2012: 79
[149] Edmondson et al. 2007: 30
[150] Tucker et al. 2007
[151] Detert and Edmondson 2011

inclusiveness, 'the words and deeds exhibited by leaders that invite and appreciate others' contributions',[152] members feel protected to talk openly about the non-conformance and induce second-order problem-solving solutions:

> Our results suggest leader inclusiveness—words and deeds by leaders that invite and appreciate others' contributions—can take nature off its course, helping to overcome status' inhibiting effects on psychological safety. In cross-disciplinary teams with high leader inclusiveness, the status-psychological safety relationship was weakened.[153]

Leaders must be active and demonstrate inclusiveness when leading a severe event.

Interdisciplinary action teams

Traditional problem-solving management research typically concentrates on the individual (e.g., how people make sense when an unplanned event occurs) or on the role of the system people operate in (e.g., normal accident theory). As the sensemaking literature on the critical nature of the empowered expert offers understanding of how people manage and treat errors, the psychology safety literature has offered a third perspective which 'integrates system and individual levels of analysis by focusing on the work group as the point where organisational and cognitive effects meet and play out in enabling or preventing errors.'[154] Psychology safety research contributes to moving to second order problem-solving and

---

[152] Nembhard and Edmondson 2006: 941
[153] Nembhard and Edmondson 2006: 958
[154] Edmondson 1996: 68

learning from mistakes by pointing to factors at the team level. As opposed to conventional problem-solving research and the focus on the training and task knowledge, psychology safety places focus on the social knowledge a team must have to perform successfully.

Severe events are complex problems and traditional teams are not sufficient for success. Interdisciplinary Action Teams (IAT) are defined as teams that must synchronise action in ambiguous, uncertain, and fast-paced situations, 'and the extent to which they are comfortable speaking up with observations, questions, and concerns may critically influence team outcomes.'[155] Typical examples are crisis response teams, emergency medical teams, and others who must work together to manage a problem in a stressful environment using expert specialised skills, and improvising and directing their activities in intense, unpredictable situations.[156] Intrinsically, based on their incomplete work content: 'Action teams face needs for coordination of member actions in uncertain situations; being interdisciplinary introduces differences in expertise and power that can threaten this coordination.'[157] Several factors have been found to determine success with IATs in alleviating the effects of severe events on psychological safety: technical expertise, touchpoints, and tacit knowledge.

---

[155] Edmondson 2003: 1419
[156] Sundstrom et al. 1990
[157] Edmondson 2003: 1444

Successful teaming: promote equality and stability to draw out technical expertise

Drawing a resemblance to individual enactment, psychological safety literature uses the verb 'teaming' rather than the noun 'team' to reflect the action oriented, flexible nature needed on complex problem-solving teams. IAT teams are temporary by definition, assembled to solve the severe event and chosen for their technical expertise: 'It's a way to gather experts in temporary groups to solve problems they're encountering for the first and perhaps only time.'[158] Teaming is about executing and learning at the same time with the challenge to 'bring together not only their own far-flung employees from various disciplines and divisions but also external specialists and stakeholders'.[159]

Teaming requires collaborative behaviour regularly at odds with big-company formal functional organisational structures, 'which divide people by specialty and focus more of their attention on bosses than on peers'.[160] Inclusive leadership is important when dealing with technical experts to prevent status differences between different functions from impeding success. When members of the IAT lack psychological safety, opportunities can be missed because of an unwillingness to engage openly with each other. Status differences can unintentionally shape the work and team environment:

> The allocation of benefits which favours high-status individuals over lower status individuals shapes the environment they share

---

[158] Edmondson 2012: 74.
[159] Edmondson 2012: 74
[160] Edmondson 2012

as well as interpersonal interactions (e.g., Alderfer 1987). Individual awareness or beliefs that membership in a particular group (e.g., profession) bestows a certain level of status creates feelings of superiority or inferiority that consistently govern behaviour so as to preserve the hierarchy.[161]

Practical considerations to minimise distractions include ensuring a culture of equanimity and equalness among the experts working on the IAT. As expense accounts in corporations often differ based on function, the inclusive leader can minimise perceived power differences by ensuring the experts eat and stay at the same lodging and using the same class of airline ticket: 'Additionally, because these choices and efforts occur within the organisational context, the organisational climate matters; psychological safety makes it all possible.'[162]

In addition, research on IAT has shown the importance of team stability. As the team of experts are learning through the problem, co-creating knowledge, team stability is a key characteristic of success.[163] And as the demands of the problem-solving effort can be strenuous and cause fatigue, data has shown that having experts on a team, by 'functioning as teams they were able to compensate for these [exhaustion], presumably because they were better able to coordinate and to catch each other's mistakes.'[164]

---

[161] Tajfel and Turner 1986; Webster and Foschi 1988 in Nembhard and Edmondson 2006: 945
[162] Tucker et al. 2007: 906
[163] Pisano et al. 2001
[164] Edmondson 1996: 71

Team touchpoints: communication needed for real time learning

The severe event problems that IAT tackle require constant and immediate communication. An IAT should react to unexpected severe events in a synchronised way, which is facilitated by unrestricted and open transfer of data and information 'to enable real-time, reciprocal coordination of action'.[165] Teams need to habitually meet to agree on action, debrief on results, and reflect in order to design any changes needed.[166] In a stark change from the structure of traditional problem-solving teams, such as 8D or Six Sigma, where lessons learned take place at the very end of the project, IAT teams conduct action and learning in parallel, investigating and drawing lessons learned from the course of action.[167] Useful not only in life-or-death situations, this is taken from the U.S. Military tool of After Action Review (AAR), which calls for the combat unit to get together immediately 'after action' to give feedback on what did and did not work in a blame-free safe discussion.[168]

Tacit knowledge: working on the edge of codified expertise

Fundamental to the success of interdisciplinary action teams is that colleagues 'know what each other know[s]'.[169] And for tacit knowledge, working on the edge of what is known, teamwork and constant communication in a fearless environment is the baseline for success.

---

[165] Edmondson 2003: 1421
[166] Pisano et al. 2001
[167] Edmondson et al. 2001
[168] Carroll and Edmondson 2002
[169] Edmondson et al. 2003

In an analysis of when traditional problem-solving fails, Tucker et al. (2002) identify two heuristics affecting success in problem-solving teams. The first is that individuals will do whatever it takes to continue and make progress on the operation to recover when a problem occurs (e.g., first-order behaviours are immediate recovery actions). The second heuristic is that individuals are inclined to keep the problem hidden and contained within their social unit or group rather than seek expert support from other functions or escalate to leadership.

Since heuristic behaviours inhibit the problem-solving effort by restricting knowledge sharing and result in a failure to bring pertinent expertise to the problem-solving effort; the problem-solving team must find ways to counteract such behaviours. In other words, explicit mechanisms are needed to ensure a problem-solving team has access to the best and most applicable knowledge about the problem at hand. These explicit mechanisms incite the opposite response of heuristic 1, getting people to open up about problems in order to reach and resolve the root cause and learn from the failure (e.g., second-order behaviours to apply continuous learning.) To accomplish this, leadership can ensure that communication structures and touchpoints are executed and engagement of staff across teams and at multiple levels is part of the process.[170] The literature reveals that for these difficult problems the team is working on the edge of codified expertise, making it vitally important that expert and tacit knowledge within the organisation is applied to the problem-solving

---

[170] Tucker and Edmondson 2002

effort. By constructing a cross-functional IAT, problem-solving heuristic 2 can be minimised.

> Tacit knowledge is difficult to transfer across sites, generally requiring individuals to accompany the knowledge ... Moreover, for improvement that relied on tacit knowledge, team composition stability was associated with faster learning.[171]

These cross functional teams bring together members' tacit knowledge on their respective subject areas of expertise. While codified knowledge can be represented by artifacts such as checklists, process maps, and work instructions, tacit knowledge brings the hands-on experience of the IAT members to the problem-solving effort.[172]

Synthesis of the lessons from Sensemaking and Psychological Safety

I took the learnings from experts in the fields of Sensemaking and Psychological Safety to design a new approach to problem-solving severe events, synthesising these lessons into actionable concepts or 'ideas to action' that can be applied by practitioners: Leadership; Design of communication structures; Empowered cross-functional experts; Enactment; Framing for Learning.

Leadership provides a safe environment and common frame the team can work from. Suitably designed communication gives sense by creating the problem formally. Having cross-functional empowered experts on the team allows for hands-on trial and error and brings valuable subject matter

---

[171] Edmondson et al. 2007, p. 7
[172] Tucker et al. 2007

knowledge to the process. Applying enactment by taking actions to create cues helps team members develop a frame. Finally, by using the frame of learning, the team moves from first-order firefighting to second-order problem-solving, creating new knowledge to learn from the failure. The ideas for action are discussed further and are then coalesced into a new approach to problem-solving that translates these ideas into practical application: T.R.U.S.T.

Idea to action 1: Leadership

In severe events, the role of the leader in problem-solving is paramount: it is an active, interlocking position that can promote an environment that supports the team in feeling safe to speak up and provide voice during a stressful time. As the severe event unfolds and employees seek sense, the leader can use framing intentionally and proactively, using enactment[173] to get the team moving and focused on action generating cues and beginning the problem-solving process. Special attention given to power distance effects can foster better outcomes, as good ideas come from many places regardless of the rank or social system.

The leader is self-effacing, humble, and aware of deficiencies and verbal in acknowledging them. The leader commits publicly to a successful outcome, risking both resources and reputation as the inclusive leader projects confidence that a resolution will be reached.

---

[173] See expanded discussion of Enactment under Mechanism 4 below.

Leaders can take specific actions to create psychological safety by reducing barriers to speaking up:

- The Leader must explicitly give permission for others to speak up: voice concerns, ask questions, and challenge assumptions.

- The Leader builds motivation by communicating the importance of what the team was selected to do and creating a shared purpose around solving the severe event.

- The Leader creates an atmosphere of information sharing: inviting input and explicitly requesting feedback and participation.

- The Leader invites others to express their concerns and models fallibility by admitting errors.

- The Leader uses coaching techniques such as providing clarification and feedback, listening to concerns, and being accessible and receptive to other ideas and questions.

Idea to action 2: Design of communication structures

Communication should be frequent and promote touch points for interacting and information sharing. Contrary to convention, meetings should be the communication vehicle of choice and should happen daily to ensure and promote making sense of the crisis. Language is the medium for rapid alignment to create shared mental models and sharing a mental model drives action. The daily meeting is a sustaining routine to unite the team. Reflection is built into the process of the daily meeting: 'what could we have done differently?' In designing a communication structure, the leader should:

- Specify points when experts assigned to the team must gather to coordinate upcoming decisions and resources.

- Identify when experts assigned to the team must analyse prior action.

- Model good, non-threatening communication to facilitate the flow of information; this must be unambiguous and understood.

- Be part of the sharing of ideas and force moments of deliberation and reflection, questioning themselves and then others in the team: 'Is this the only way to see the situation? What might I be missing?' What could we have done differently with what we know now?'

- Provide sense through communication to the greater organisation. The communication structure should define how and when to communicate upwards, horizontally, and downwards.

Idea to action 3: Empowered cross-functional technical experts

A team should be assembled based on technical skills and expertise and not on the organisation's hierarchy. When encountering a severe event, the team is working on the edge of codified knowledge and having experts with tacit competence promotes team enactment through action. The team is empowered by the leadership providing a safe environment because excessive pressure can turn an expert into an 'advanced beginner'. The leader can facilitate the empowerment of cross-functional technical experts by:

- Hiring experts for the cross-functional team and assigning them for the duration of the event.

- Minimising status differences between team members, for example, by travelling in the same class accommodations and by eating together.

- Sending the team to the source of the crisis and being on the ground throughout the event as much as possible.

Idea to action 4: Enactment – learning by doing

A leader and the empowered team of experts do not make decisions during a severe event. A decision is something to be polished and protected; during a crisis, it is more important to make sense. Enactment is creating new knowledge through actions. Two abductive questions guide the team's behaviour: 'what is the story?' and 'now what?' The answers provide tangible cues to discovery, linking thinking and action to learn and solve the problem.

- The leader supports the team of experts to accomplish enactment through experimentation and trial and error: each experiment creates the knowledge that will be used to determine the root cause and resolve the severe event.

- The leader is flexible and protects the team from 'cartesian anxiety': the world is not black and white, and the team will have to be flexible in making mistakes.

- The leader applies the philosophy of fail fast: if there is a mistake to be made, make it quickly to enable learning and more action.

Idea to action 5: Framing for learning - Social Construction of Reality

How a problem is framed can make the challenge of a severe event compelling and career defining rather than threatening and painful. Framing is a tool for psychological interpretation the leader utilises implicitly to communicate and make sense as people react to events. By socially constructing and talking a mental model into existence, the leader can filter the environment. These mental paradigms affect the choices and decisions the team makes by defining the boundaries. Using frames, leaders can directly impact the behaviours of the team and other stakeholders in the organisation.

- The leader must communicate that the severe event is different from typical problems and that due to its severity, grants an opportunity to generate new knowledge.

- The leader communicates that due to its time-sensitive nature, this problem is critical to the organisation and its customers and support is needed on all levels.

- The leader uses face-to-face communication to reinforce how vitally important the expert is to a successful outcome. 'We are unable to achieve without the team and without the willing participation of others'.

- The leader stresses the problem will not be easy, but it will be worth it as new skills and experience will be created from successfully solving the problem.

- The leader communicates that the team is vitally important to success and may bring key pieces of the puzzle that cannot be anticipated in advance. 'We are in it together and will make it happen together'.

CHAPTER 5 T.R.U.S.T. for Practitioners, Step-by-Step

The T.R.U.S.T. approach synthesises ideas from academic literature and expert practitioners into a practical plan of action. T.R.U.S.T. includes five operating principles and six tools to guide the practitioner applying the solution created by a team of cross-functional experts that work collaboratively to problem-solve a S.H.I.T. problem by identifying and formalising second-order problem-solving, i.e., lessons learned and preventive actions. The T.R.U.S.T. approach is designed to foster an outcome where the 3P, i.e., Pressure, People and Political effects, are mitigated, enabling successful problem-solving efforts that effectively resolve the problem and reduce the negative psychological impacts of the problem on company personnel. Figure 15 outlines how T.R.U.S.T. works.

Challenge
The social mechanisms of People, Pressure, and Politics (3P) hijack and sabotage the problem-solving process when S.H.I.T. conditions are present. Second order problem-solving learning is missed.

Approach to Overcoming the Challenge
T.R.U.S.T. solution brings together inclusive leadership, psychological safety, collaborative team environment, enhanced communication structures, knowledge sharing, empowered cross-functional experts, enactment, social constructions, and framing to overcome the effects of 3P.

Outcomes
3P effects are mitigated through creation of an empowered team of cross-functional experts who work collaboratively to successfully resolve the problem and identify second order problem-solving.

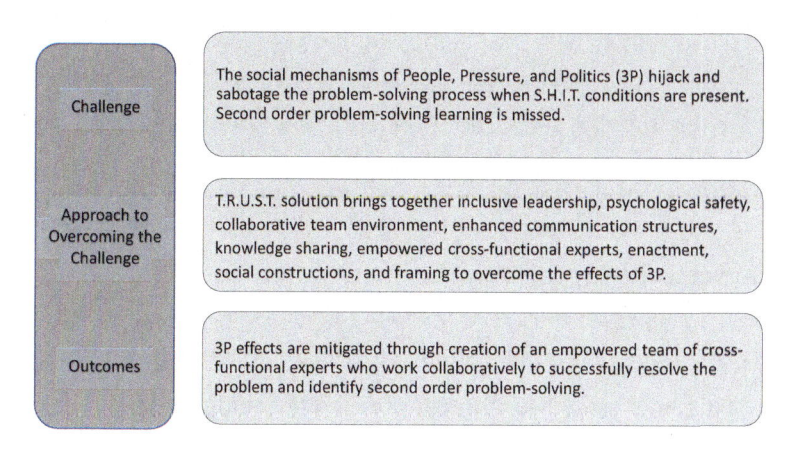

Figure 15   How T.R.U.S.T. Works

The T.R.U.S.T. solution incorporates mechanisms established by academic thought-leaders in behavioural science to augment traditional problem-solving, so that practitioners are equipped to mitigate and move through the human behaviours that cause problem-solving efforts to fail. The simple mnemonic "T.R.U.S.T." is a framework for communicating the operating principles of the solution:

(T) Tsar is appointed to act independently to govern and support the problem-solving team. The Tsar promotes psychological safety and initiative sensemaking through formal acknowledgment of personal limitations and creation of an environment where blame is not tolerated and speaking up/dissent is safe.

(R) Re-frame failure as an opportunity for knowledge. As the team is working on the edge of codified knowledge, there is no manual or design guide. By methodically performing design of experiment with trial and error to collect, analyse, store, disseminate, and learn, the team can treat each failed experiment as a lesson, moving it closer to a solution.

(U) Understand and define the problem. The problem is a social construction. The Tsar provides the team and organisation sense and awareness by taking ownership of the problem, defining its boundaries, and exploring multiple perspectives to frame the problem to be solved.

(S) Support open communication structures. Establish continuous communication of real-time information about the context of the problem and the actions taken. Communication is truthful, unambiguous, and challengeable by the team.

(T) Technical team of empowered experts is formed. A problem-solving team of technical experts with relevant specialisation including cross-functional actors and representing a 'team of teams' helps to interpret the scientific knowledge of multiple organisational functions.

Six tools have been designed from the mechanisms identified in literature that will be used for the intervention to support the outcome: identification and successful resolution of the S.H.I.T. problem.

First is an evaluation tool to identify when S.H.I.T. conditions exist. Once it is determined that a S.H.I.T. problem is underway, the rest of the tools support leaders and teams in applying the new, augmented approach to problem-solving. The tools are:

1. S.H.I.T. Problem-Solving Identification Evaluation
2. Problem-Solving Group Model to Build T.R.U.S.T.
3. Situational Analysis: Frame the Problem
4. Severe Event Alert
5. Permission Charter to Promote Psychological Safety
6. T.R.U.S.T. In Action Checklist

Each tool is available for download at theshitshow.biz and is designed so that firms may customise as needed for their environment. The following sections describe each tool and provide its usage specifications, briefly describing the purpose of the tool, the roles and responsibilities of personnel, the process for using the tool, and an overview of the tool itself.

Tool 1. S.H.I.T. Problem-Solving Identification Evaluation

The first step of the solution is to identify and classify the situation as a S.H.I.T. problem. This will signal that traditional problem-solving methods need to be augmented to mitigate the 3P of Pressure, People, and Politics, move beyond first-order problem-solving behaviour to second-order root cause analysis, and promote continuous learning. This tool defines the process of classifying the problem as a S.H.I.T. problem.

Roles and Responsibilities

It is the responsibility of the organisation's Quality Manager to complete the template using the available information and to communicate the results to senior business management.

Smaller firms often combine the role of Quality Manager with that of the Operations Manager. In this case, whoever is the appointed representative of the management system would complete the template.

Process

The organisation's Quality Manager completes the S.H.I.T. Problem-Solving Identification Evaluation template. The completed template is then distributed via email. As the full impact of a severe event is unknown and potentially affects the whole organisation, the mandatory distribution list should include all senior leadership of the firm.

Senior leadership can vary based on the size of a firm and organisation. A guideline principle for the appropriate senior leadership is the respective P&L (profit and loss) owner of the business segment. Typically, this may be a product line, division, or business unit structure.

The templates should be stored in a centralised location as a reference for sharing and learning. The process for completing the template and disseminating the results is outlined in Figure 16.

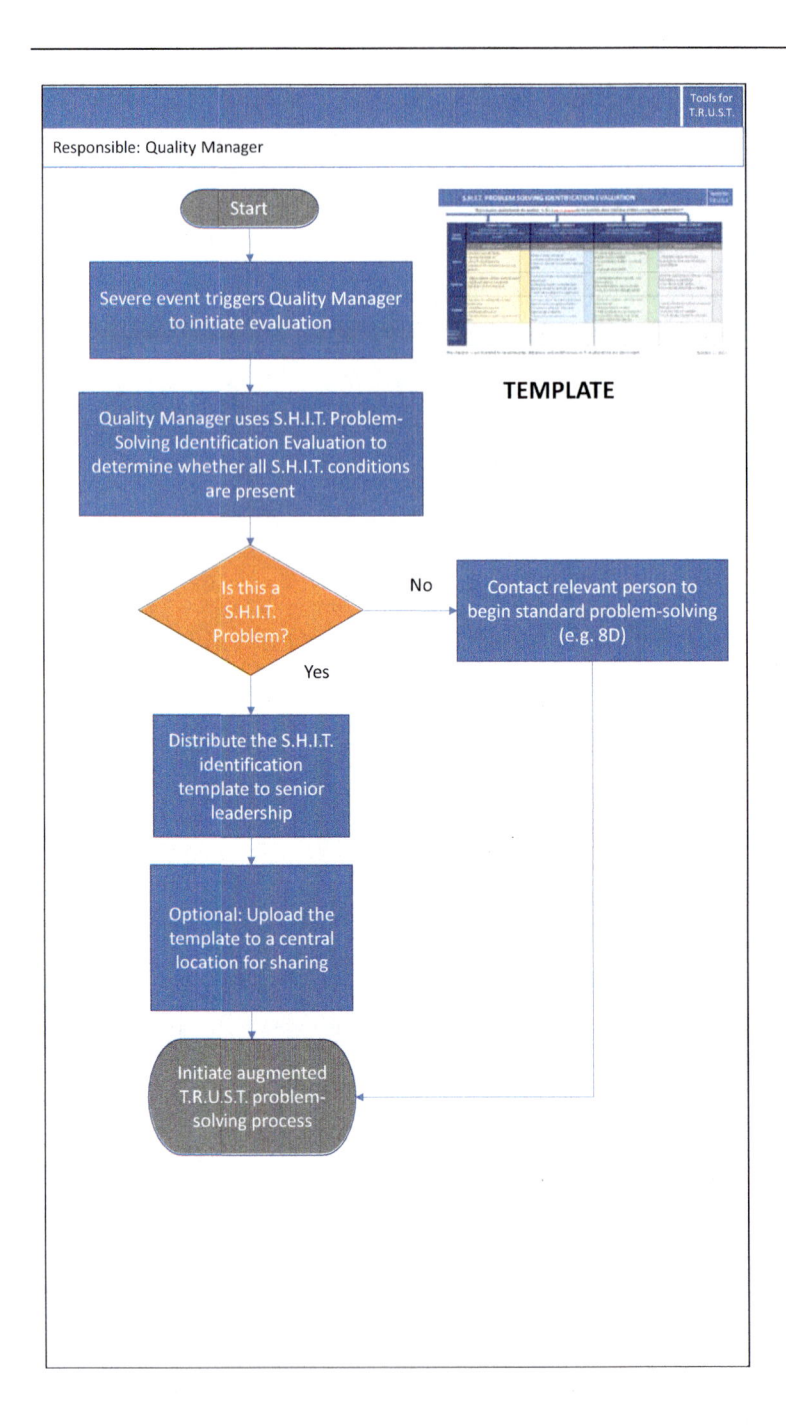

Figure 16 S.H.I.T. Problem Identification Process

Tool Overview

The template for the S.H.I.T. Problem-Solving Identification Evaluation (Figure 17) lists the elements of a S.H.I.T. problem (Severe Event, Highly Visible, Information Deficient, Time-Sensitive) and a categorisation of the problem according to system difficulty (routine, significant, or complex).[174]

For smaller firms, local customisation could be to combine routine and significant into one category to simplify the template and process.

If either the significant or complex row is checked for all four of the S.H.I.T. elements, then the results from the completed template are to be interpreted as a 'yes' and this problem should be classified as a S.H.I.T. problem.

---

[174] A pdf version of this template is downloadable for free at: https://theshitshow.biz/tools

Figure 17   S.H.I.T. Problem-Solving Identification Evaluation

S.H.I.T. PROBLEM SOLVING IDENTIFICATION EVALUATION

Tools for T.R.U.S.T.

This evaluation should answer the question, "Is this a S.H.I.T problem for the business where traditional problem-solving needs augmentation?"

System Difficulty	Severe Event? *Is the impact severe to your business? Is the event impacting Health and Safety as well as Customer?*		Highly Visible? *Is the event visible in your internal company? Is the company in the news or other multi-media?*		Information Deficient? *Do you need to secure information? Can the interaces be managed and controlled?*		Time-Critical? *Are there external or internal time constraints? Is the end Customer impacted?*	
	Minimum Criteria	Check	Minimum Criteria	Check	Minimum Criteria	Check	Minimum Criteria	Check
Routine	– Minimal Customer impact – Minimal Technical risk – Minor Financial exposure – Operations not impacted (no stop work required)		– Event is locally contained – Customer involvement not required – Internal Customer involvement / approval needed		– Problem repetitive and solution is readily available (no RCA needed) – No investigation needed - root cause known – Local expertise available		– Achievable in given time frame. No actual internal or external customer impact affected	
Significant	– High probability of major technical issues – Significant repair work expected – Significant Financial exposure		– Customer escape - non-conformance left value stream – External Customer / authorized body approval needed for technical solution – Event highly visible in the organization		– Engineering interface required - root cause unknown – Multiple locations may be involved – Product or Process redesign needed		– External customer order affected, overall time schedule is jeopardized – Crisis recovery plan needed – Customer penalties a high-probability	
Complex	– Certainty for replacement of major components – Immediate stop required – Confirmed safety issue – Potential product recall - warranty event likely		– Company loss of reputation threatened - external media management needed – Customer involvement - 3rd party litigation high-probability – Serial production impacted / product recall		– Technical interfaces - cross functional team required – Multiple locations involved – Product redesign driving Production / Service process redesign (e.g. Design concept / technology redesign)		– Impact of external customer contractual dates jeopardized – Customer stop work enacted – Liquid damages / penalties actualized	
Comment / Reason for selection								

This checklist is not intended to be exhaustive. Adaptions and modifications to fit local practice are encouraged.

Revision 1 / 2022

Severe Event. The guiding question for a severe event is the impact on the welfare, health, and safety of the employees, the customers, and the business. The guidance criteria are intentionally qualitative and normative without financial measures, as each business implementing this process will have local conditions on the significance or added complexity of a monetary or commercial penalty. For example, a quality executive from an American multinational stated that internal metrics for a severe event were more than $500,000 per incident. At his German competitor, the measure was how many units could be affected in the fleet or in production, as end-user impact was viewed as more severe than financial penalty.

Highly Visible. The guiding questions for a highly visible event consist of internal, intrafirm, and external coverage in the media. In addition, added complexity for a visible event can include whether an external customer needs to be involved and if a legal action, such as a warranty event, is applicable and local or national government would need to be notified (e.g., U.S. Consumer Product Safety Commission).

Information Deficient. The guiding questions for an information deficient event include the potential impact of cross-functional technical interfaces (each function with their own specialised jargon) and the availability of information on the scientific details. A problem is routine when no additional expertise is needed; a problem is complex when it involves engineering on the edge of codified knowledge and experts are needed to design experiments and co-create technical specifications as part of the problem-solving. Often, with a complex problem, the production process and the product need a redesign in parallel, adding to the complexity of the second-order root cause analysis. In addition, with the event of a product recall the team may need to also design a service solution for the fleet of products already sold.

Time-Sensitive. The guiding questions for a time-sensitive event include project or production time commitments, worsening consequences or deteriorating conditions as time passes, and the possibility of legal and/or financial penalties that increase daily. An example is when a power station is off-line, or when there is a daily penalty for missing parts in a manufacturing process. In addition, many nations have laws and/or regulations that specify the time allowed for reporting a severe event if user harm is possible due to product failure (e.g., the U.S. Consumer Product Safety Commission).

Tool 2. Problem-Solving Model to Build T.R.U.S.T.

Once the problem has been classified as a S.H.I.T. problem, the leadership team should appoint an independent Tsar with appropriate authority to manage the S.H.I.T. problem-solving effort. The problem-solving group model outlines the way the Tsar should assemble the group, how the group should function, and how it should conduct a situational analysis.

Roles and Responsibilities

It is the responsibility of the organisation's senior management team to appoint a Tsar who will be empowered to lead the organisation through the S.H.I.T. problem. The Problem Tsar does not have to be a full-time role; this may vary based on the local firm environment.

The Tsar will assemble a group of cross-functional experts from different areas of the company who are fully assigned to the problem-solving team. The Problem Tsar is the independent leader of the team, and is not appointed as a technical expert who drives the technical decisions but instead supports the team to make sense of the problem and ensures continued action and interpretation.

The Problem Tsar promotes psychological safety and makes the team feel protected from organisational politics. During the introductory meeting, the Tsar formally acknowledges the limitations and the independent nature of the team.

The Problem Tsar will establish, at minimum, daily communication with the team to synthesise the evolving context into a coherent story which can be communicated weekly to the senior management and key stakeholders.

The Problem Tsar will initiate a social media chat group for real-time communication for all the Technical Experts involved in the S.H.I.T. problem-solving efforts. Examples of real-time communication applications available at the time of this writing include WeChat, (especially mainland China), WhatsApp, Microsoft Teams, Line, etc.

The technical experts (TE) are appointed to the S.H.I.T. problem-solving team. The TE are drawn from functional domains such as research and development, engineering, manufacturing, procurement and quality. The group members shall represent their function and escalate and draw on the resources from their function, as necessary. They are the representatives of their functions, responsible for identifying resources to facilitate action and progress through investigation to co-create new knowledge. As the

S.H.I.T. problem involves working at the edge of codified and documented knowledge, the TE group uses trial and error to generate new knowledge for the root cause analysis.

If possible, the function leader should be a member of the daily communication report to ensure sufficient support and remove roadblocks as they may occur. Examples would be the head of engineering or the head of quality.

Process

Once the S.H.I.T. problem is confirmed, the leader appoints a Problem Tsar. The Problem Tsar is then directed to assemble a team of technical experts.

The process flow chart is shown in figure 18. A template summarising the operating principles and visualisation of the problem-solving group follows.

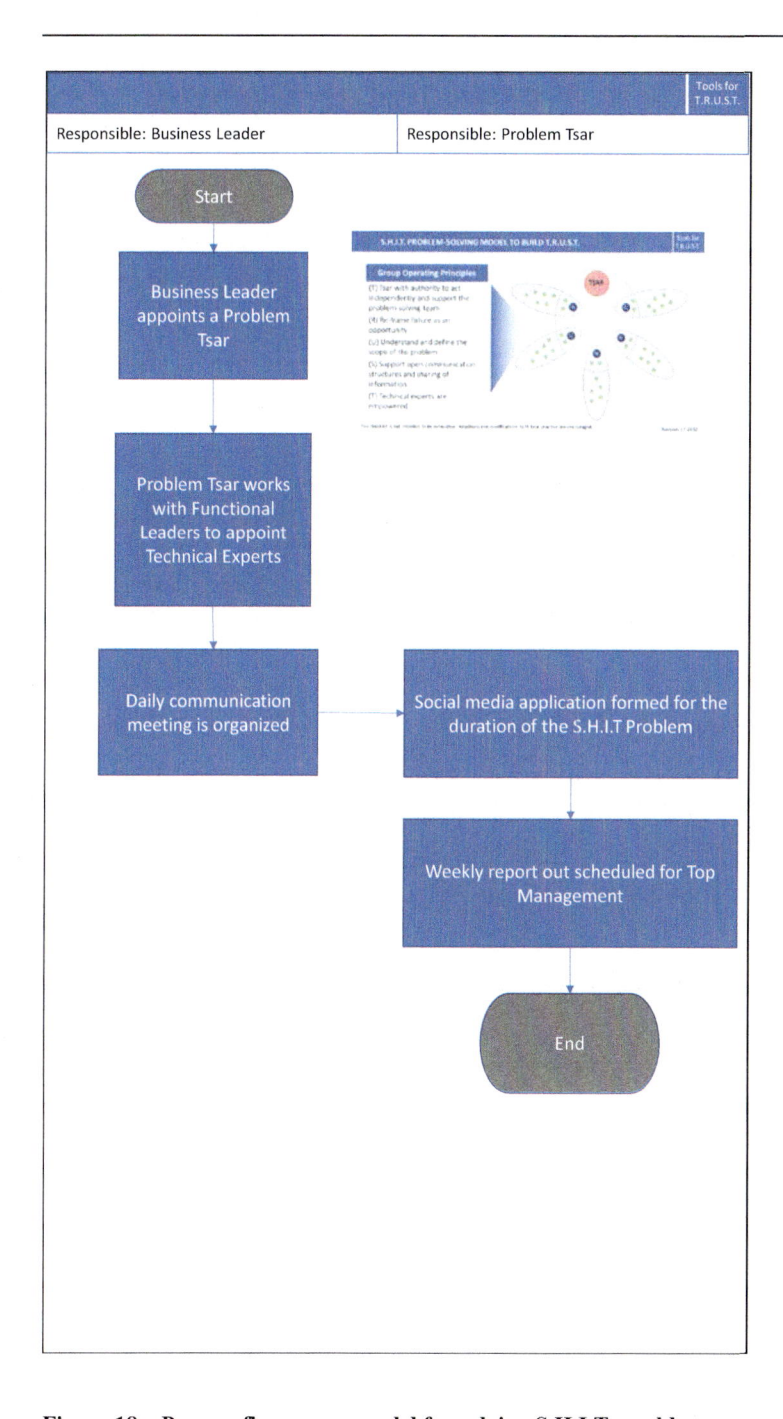

Figure 18 Process flow group model for solving S.H.I.T. problems

Tool Overview

The template 'S.H.I.T. Problem Solving Model to Build T.R.U.S.T.' (figure 19) consists of a visualisation of the S.H.I.T. problem-solving group and the operating principles on how the group is organised and led.

Figure 19 Problem-Solving Model to Build T.R.U.S.T.

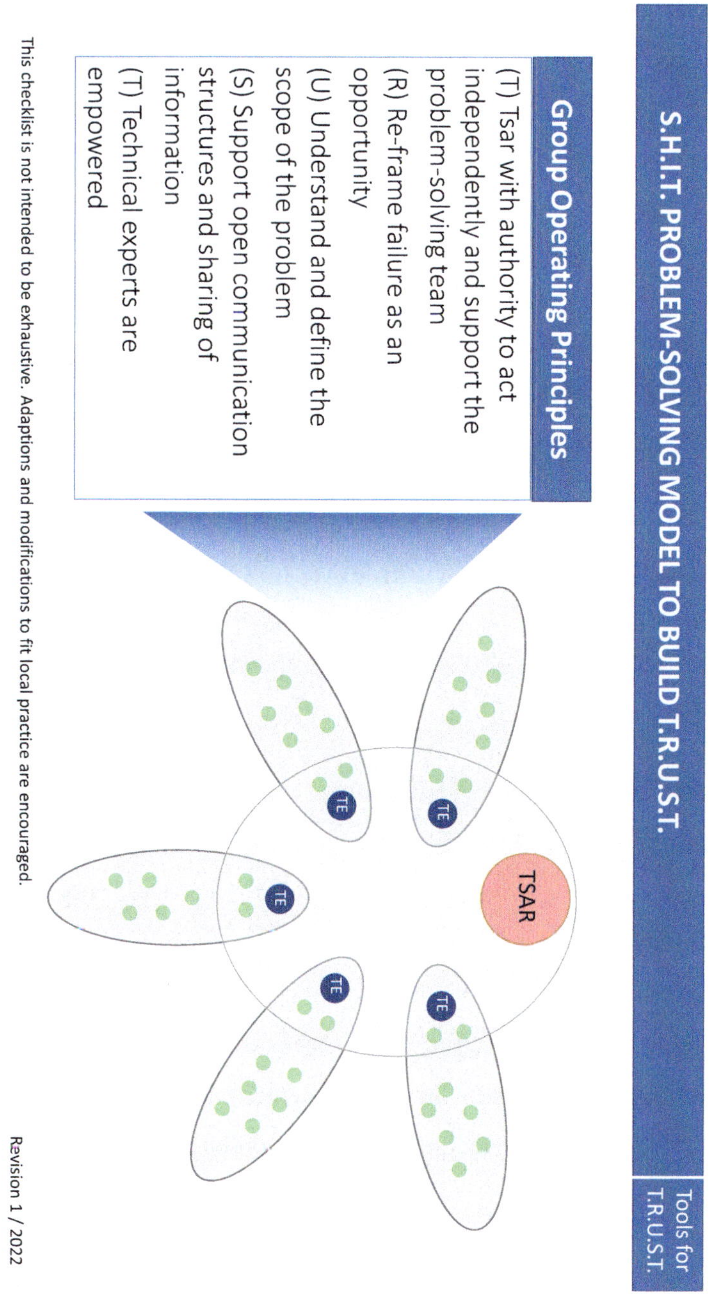
S.H.I.T. PROBLEM-SOLVING MODEL TO BUILD T.R.U.S.T.
Tools for T.R.U.S.T.
Group Operating Principles
(T) Tsar with authority to act independently and support the problem-solving team
(R) Re-frame failure as an opportunity
(U) Understand and define the scope of the problem
(S) Support open communication structures and sharing of information
(T) Technical experts are empowered
TSAR
TE
TE
TE
TE
TE
This checklist is not intended to be exhaustive. Adaptions and modifications to fit local practice are encouraged.
Revision 1 / 2022

Organisation of the S.H.I.T. Problem-Solving Group. As visualised in the template (figure 19) 'Problem Solving Model to Build T.R.U.S.T.', the S.H.I.T. problem-solving team is a group of technical professionals chosen for their expertise as opposed to traditional problem-solving skills or organisational rank. The Tsar is not superior to the group as a decision-making authority, instead the Tsar is an executive to remove roadblocks, synthesise information, and facilitate consistent real-time communication.

The team works to address different lines of inquiry simultaneously and in parallel. The technical experts are the representatives of relevant functions, responsible for bringing resources necessary to take action and further the investigation through trial and error to co-create new knowledge.

Group Operating Principles – Model to build T.R.U.S.T.

<u>(T) Tsar Appointed Independently to govern and support the problem-solving team</u>

The Tsar should be as independent as possible. The extent to which independence is attainable will vary depending on a variety of factors such as the size of the organisation and the structure. For larger organisations, typical roles which by the nature of their function have independence at their core include quality, health and safety, and operational excellence. Independence facilitates getting to the root cause and driving improvement as it mitigates personal motivations that might inhibit the process, enabling a greater focus on understanding rather than outcome.

The Tsar will also support the group by managing upward communication.

(R) Re-frame failure as an opportunity for knowledge

The Tsar takes measures to reduce team stress and provide a psychologically safe space. The Tsar achieves this through a variety of actions including: creating an inclusive environment empowering TE to make decisions based on their experience, supporting the team to make sense, and removing blame while setting an expectation that failure is part of learning, which is the desired outcome. It is safe to make mistakes in this group as information is not fully formed and is generated during actions.

(U) Understand and define the scope of the problem

The Tsar will articulate a clear problem statement and direct the team to understand the problem from different perspectives, inviting multiple perspectives from the problem-solving team. In a S.H.I.T. problem, information is deficient. By driving trial and error with rapid failures and quick lessons, the team can gain knowledge to better understand the scope and the characteristics of the problem, and the problem's symptoms and their effects.

(S) Support open communication structures and sharing of information

Active, frequent communication is the glue for the technical experts working on a S.H.I.T. problem. The environment is complex, and information is created by taking actions (recall the concept of 'enactment' in sensemaking). Frequent touch points and a variety of communication media (i.e., chat, social networking, IT, etc.) can help ensure equal access to information.

(T) Technical Experts empowered to discover

Cross-functional Technical Experts are empowered to learn and discover the root cause wherever it may be. The creation of a 'no-blame environment' facilitates this inquiry by providing a psychologically safe space. The team creates knowledge through trial and error to further understanding and uncover the root cause of the problem.

Tool 3. Situational Analysis – Frame the Problem – 5W2H

After the problem has been classified as a S.H.I.T. problem, it must be framed and defined. The Problem Tsar is appointed and begins the communication process, such as officially delimiting the boundaries of the problem, creating an organisation alert for enterprise communication (the next tool), and performing a situational analysis. An established method for formulating the situational analysis that is complementary to our augmented problem-solving approach is the 5W2H method.[175] Though this method is well-known, it's currently employed on an ad-hoc basis. Incorporating it into the technical problem-solving process will provide a familiar approach to achieve the sensemaking that is needed in a S.H.I.T. situation. The template provided includes questions designed to systematically analyse the cues from the environment which will aid the team to formally create the specification of the problem.

---

[175] The 5W2H (what, why, where, when, who, how much, how often) method is a well-known process for organising thinking around problem-solving. Further information is available through many additional resources from the American Society for Quality and its European and Asian counterparts.

Roles and Responsibilities

The Problem Tsar convenes a meeting of the cross-functional technical experts to answer the questions from the 5W2H template. The Problem Tsar then combines the group's answers to create a coherent story which provides the narrative of the severe event. The technical experts will contribute to the 5W2H process by applying their knowledge to analyse the cues from the environment context and answer the questions to formulate the situational analysis.

Process

A situational analysis is conducted using the 5W2H tool in order to formally create the problem. The Tsar begins by convening the team and calling for the 5W2H analysis. The team of technical experts then completes the 5W2H analysis by applying their expertise and answering the questions. The team provides the results to the Problem Tsar, who then synthesises the results to create the narrative of the severe event.

Tool Overview

Figure 20 lists the steps that must occur as well as the roles and responsibilities of the Problem Tsar and the team of technical experts.

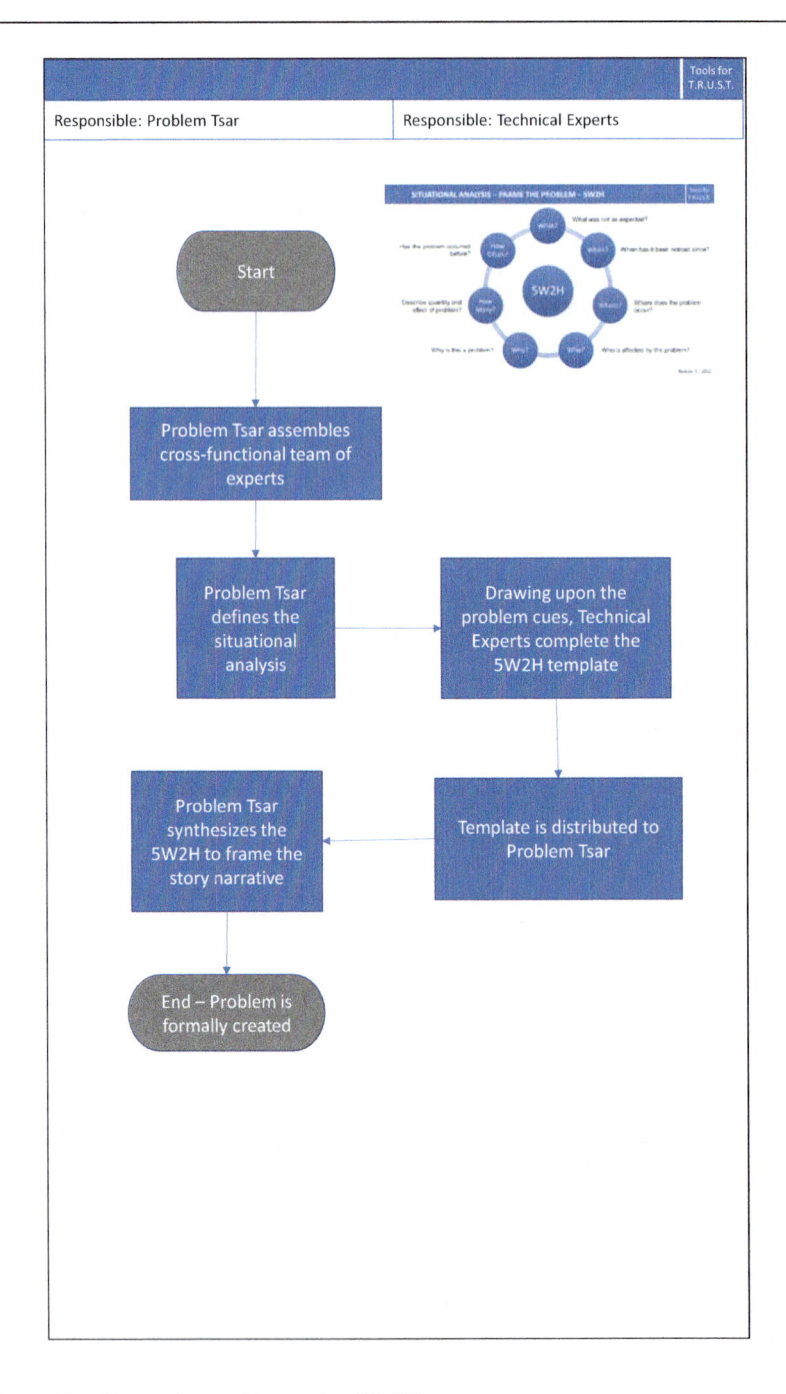

Figure 20   Frame the problem using 5W2H, step-by-step

The situational analysis 5W2H Template (figure 21) consists of a series of questions to prompt the team to systematically analyse the cues from the environment to frame the problem, develop common group understanding, and facilitate learning. The questions included in figure 21 are designed to prompt thorough and thoughtful consideration of the variables surrounding the problem, but they are not exhaustive. Dependent on the problem characteristics, the 5W2H process may also include consideration of a variety of additional questions. Some examples are:

What?

- What happened that created the S.H.I.T. conditions?
- What happened that was not expected?
- What products are affected?
- What are the consequences of the event? If this has happened before, what was the previous conclusion and corrections?

Who?

- Who is affected by the event?
- Who first observed the event (internal/external)? Which department or functional area?
- Who did not find the event in previous process/activity?

Why?

- Why did this happen?
- Why is this a problem?
- Why aren't there processes in place to prevent an event like this?

Where?

- Where was the event observed (process, location, area, department, site)?
- Where does the event occur? Where does the event not occur?

How Much/Many?

- Describe quantity and effect of the event?
- How is the event impacting operations measured in money, people, and time?
- How much impact could the event have in other ways, beyond money, people, and time?
- How many units could have the problem?
- What impacts are we unable to quantify?
- How large could the impact be?

How Often?

- Has the event occurred previously?
- How often has this event occurred?
- What is the trend (continuous, random, periodical)?
- How pervasive is the impact of the event over time (number of non-conformities, performance indicators not fulfilled)?

SITUATIONAL ANALYSIS – FRAME THE PROBLEM – 5W2H *Tools for T.R.U.S.T.*

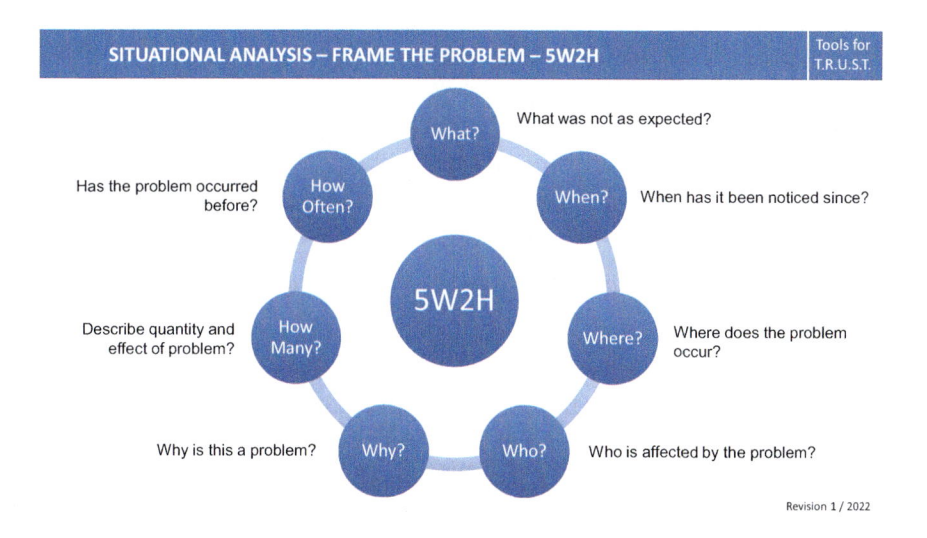

- What? — What was not as expected?
- When? — When has it been noticed since?
- Where? — Where does the problem occur?
- Who? — Who is affected by the problem?
- Why? — Why is this a problem?
- How Many? — Describe quantity and effect of problem?
- How Often? — Has the problem occurred before?

5W2H

Revision 1 / 2022

Figure 21   Template for Situational Analysis Using 5W2H[176]

Tool 4. Severe Event Alert

The Severe Event Alert is a communication vehicle that brings the organisation into alignment by officially recognising the event and placing boundaries on the scope of the problem. Once the official alert is distributed, the S.H.I.T. problem is real, and the organisation is aware of the severity of the event and a need for action.

Roles and Responsibilities

After the narrative of the problem is developed through the 5W2H process, the Problem Tsar is responsible for distributing the Severe Event Alert to all affected stakeholders of the event. Ideally, organisations adopting this augmented approach will have identified a distribution list for severe event

---

[176] A pdf version of this template is downloadable for free at: https://theshitshow.biz/tools

notification. At a minimum, all members of the senior leadership team, including functional leaders, must be included. It can also be good practice to share the Severe Event Alert with customers to demonstrate recognition of event severity and the action the company is undertaking.

The technical experts collect information on the severe event and communicate the situational analysis to the Problem Tsar to populate the alert template.

Process

S.H.I.T. conditions have been confirmed and a 5W2H analysis completed. The Problem Tsar takes this information to create the event narrative. Working with the team of experts, the Tsar completes the Severe Event Alert template and circulates the alert via email as a PDF file. Even if the company has an online automatic alert database, the S.H.I.T. alert is personally distributed.

Tool Overview

The 'Severe Event Alert Template' (Figure 23) consists of four areas. Instructions for completing the template follow.

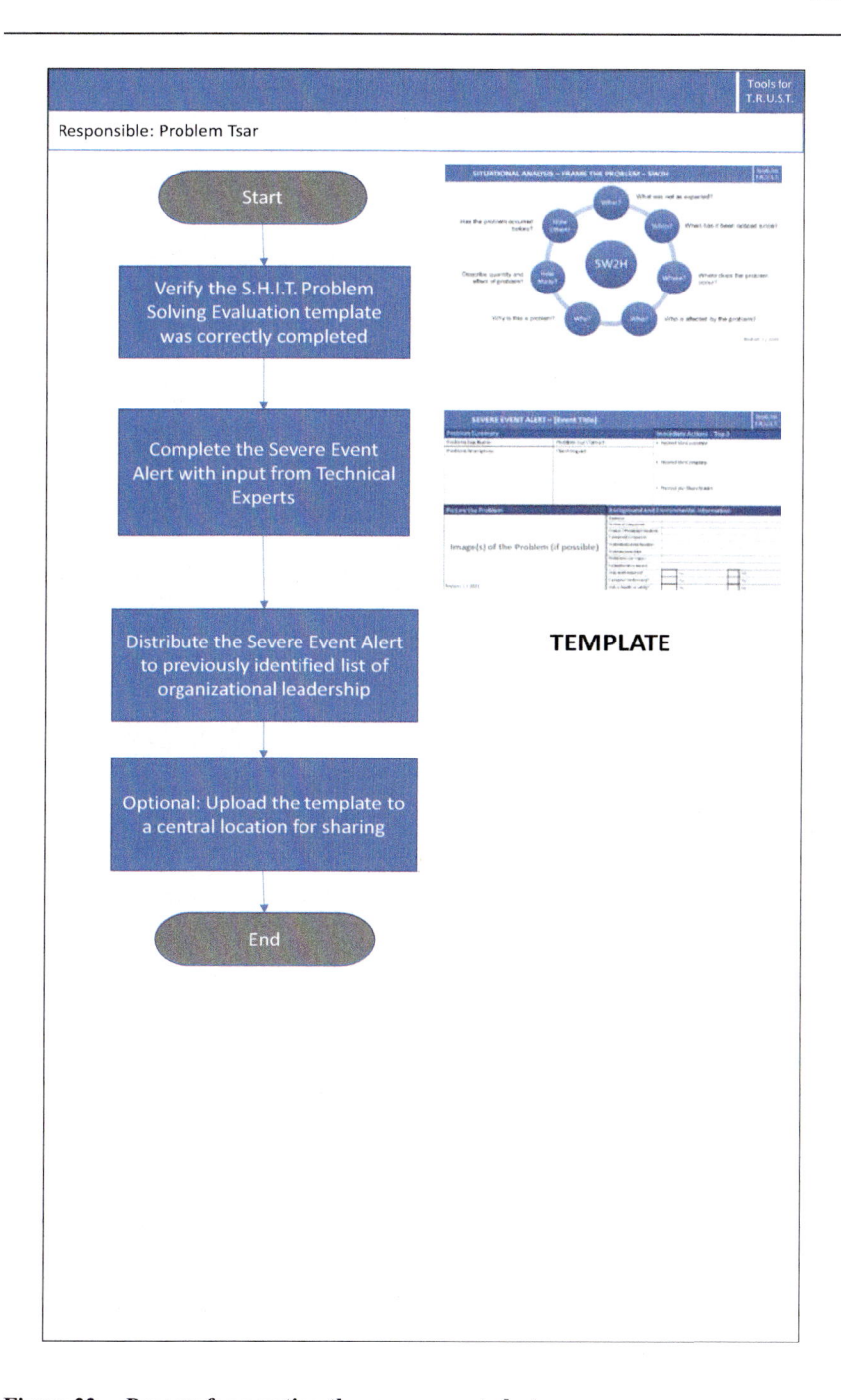

Figure 22 Process for creating the severe event alert

150

Figure 23 Severe Event Alert Template

SEVERE EVENT ALERT – [Event Title]

Tools for T.R.U.S.T.

Problem Summary

Problem Tsar Name:	Problem Tsar Contact:
Problem Description:	Client Impact:

Immediate Actions - Top 3

- *Protect the Customer*
- *Protect the Company*
- *Protect the Shareholder*

Picture the Problem

Image(s) of the Problem (if possible)

Background and Environmental Information

Business:	
Technical component:	
Project / Production location:	
Customer(s) impacted:	
Problem detected location:	
Problem issue date:	
Estimated cost impact:	
Estimated time impact:	
Stop work required?	Yes / No
Customer involvement?	Yes / No
Risk to health or safety?	Yes / No

Revision 1 / 2022

Instructions

Problem Summary: Fill in the name and contact information (phone and email) for the Problem Tsar. Next, describe the problem using information from the 5W2H situational analysis. In the Client Impact box, describe what the customer or client experienced when the problem was detected and/or occurred, if applicable. This descriptive language should identify the customer or clients impacted and succinctly describe the problem.

Immediate Actions, Top 3: Outline immediate actions taken to minimise the risk and impact to the customer, the company, and the shareholders associated with the problem. For example, the customer may need to deactivate equipment or limit its use. Other sites within the company may need to be alerted and inventory checked.

Visualise the Problem: Use this area to insert picture(s) or sketch (es) of the problem.

Background and Environmental Information: Complete this section using information identified from the work of the problem-solving team. Include the business responsible for the problem-solving management; the system, component, and/or part which the problem items belong to; the name of the project or manufacturing site where the problem occurred; the type of site where the nonconformance was detected (i.e., factory, external supplier, engineering); the names of customers impacted by the problem; the date when the problem was first recorded; the estimated cost to the firm; and the number of weeks the problem will likely delay the critical path. For the yes/no boxes at the bottom of this section, mark an "X" answering whether the problem has led to work stoppage at the customer site or at the plant or project; whether or not final approval is required from the customer for the correction of the problem; and whether or not the health and safety of personnel are at risk.

Tool 5. Permission Charter to Promote Psychological Safety

The permission charter is a document from the Problem Tsar to the technical experts and other key stakeholders communicating that the team is empowered to solve the problem. The objective of the permission charter is to help ensure that the Problem Tsar demonstrates inclusive leadership, lowers the perceived cost of speaking up, and removes roadblocks to create a blame-free environment while ensuring consistent communication.

Roles and Responsibilities

The Problem Tsar is responsible for completing and distributing the permission charter to all affected stakeholders of the S.H.I.T. problem-solving team.

Process

Once the Severe Event Alert is shared. The Problem Tsar turns to empowerment of the team by creating a safe environment where making mistakes is expected and tolerated. This is formally communicated via a permission charter shared during the kick-off meeting. The message is also verbally communicated, and the team is invited to provide feedback.

Tool Overview

The template 'Permission Charter to Promote Psychological Safety' (figure 24) consists of three areas.

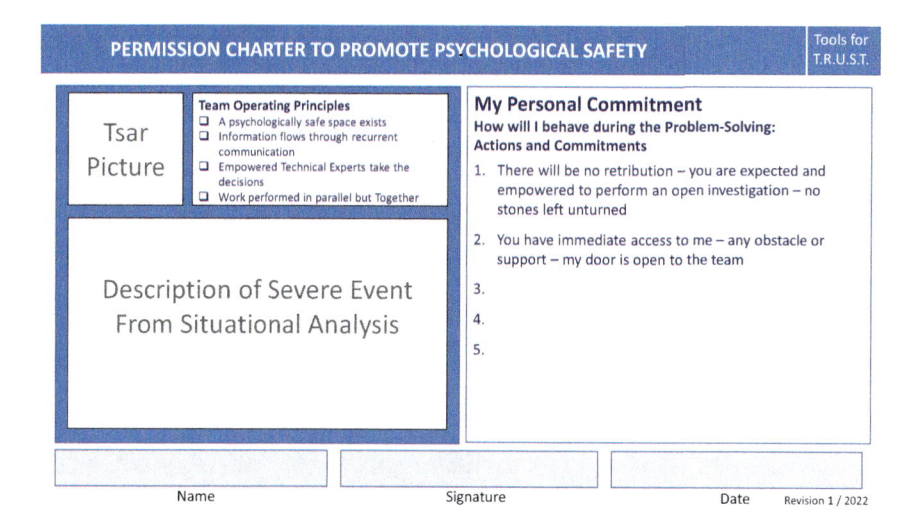

Figure 24   Permission Charter to Promote Psychological Safety

Tsar Picture. This document emphasises a commitment to a blame-free environment. The Problem Tsar includes a photo on the document to signify personal commitment to the team and ownership of the problem.

Description of Severe Event from Situational Analysis. Overcommunication is a feature of the T.R.U.S.T. problem-solving approach. The Problem Tsar will restate the scope and boundaries of the problem from the 5W2H situational analysis, summarising the story narrative.

My Personal Commitment. The Problem Tsar outlines a personal commitment to the team for a psychologically safe working environment where mistakes are tolerated and encouraged to promote action and learning during the root cause analysis.

Tool 6. T.R.U.S.T In Action Checklist

Inspired by the WHO Surgical Safety Checklist, the T.R.U.S.T. In Action Checklist is a document designed to enable the Problem Tsar to support the team throughout the problem-solving process. It encompasses the other templates included in the augmented solution set for S.H.I.T. problem-solving efforts and helps to ensure that processes and behaviours create a psychologically safe environment.

Roles and Responsibilities

The Problem Tsar is responsible for completing and distributing the checklist to all affected stakeholders including the technical experts on the S.H.I.T. problem-solving team. The Problem Tsar shall communicate the checklist during the kick-off meeting with the technical experts and verbally communicate and invite team feedback.

Process

The T.R.U.S.T. In Action Checklist aligns the team and formalises the actions and required elements of the augmented S.H.I.T. problem-solving approach including:

Action	Actor
Identification and confirmation that S.H.I.T. conditions are present	Business Leader
Appointment of a Problem Tsar	Business Leader
Assembly of a problem-solving team of cross-functional technical experts	Problem Tsar
Completion of a 5W2H analysis	Team of cross-functional technical experts
Synthesises of the results to create the event narrative	Problem Tsar
Formal communication of the narrative and framing of the problem using the Severe Event Alert	Problem Tsar
Creation of a permission charter explicitly communicating to the team that they are working in a safe environment where making mistakes is expected and tolerated	Problem Tsar

Tool Overview

The template 'T.R.U.S.T. In Action Checklist' (figure 25) prompts the completion of three steps: creation of the problem, assembling the group of experts, and taking action.

Figure 25 T.R.U.S.T. In Action Checklist

T.R.U.S.T. IN ACTION CHECKLIST

Tools for T.R.U.S.T.

SITUATIONAL ANALYSIS — Create the Problem

- Confirm Tsar has cross-functional authority
  ☐ Yes
- Confirm Tsar completes 5W2H Situational Analysis
  ☐ Yes
- Confirm Tsar officially communicates findings from S.H.I.T. Problem-Solving Identification Evaluation to relevant senior personnel
  ☐ Yes
- Confirm Severe Event alert distributed to all relevant internal stakeholders
  ☐ Yes
- Confirm cross-functional group of experts appointed to team
  ☐ Yes
- Confirm rapid communication tools created (e.g. messaging group, daily meetings....)
  ☐ Yes

MAKE SAFE — Assemble Group of Experts

- Confirm all group members have introduced themselves by name and role – cross-functional team of teams from relevant functional areas
  ☐ Yes
- Confirm Tsar completes permission charter to promote a safe environment and verbally communicates permission to speak openly
  ☐ Yes
- Confirm Tsar reframes failure as opportunity for knowledge and promotes continuous learning through trial and error
  ☐ Yes
- Confirm decisions lie with technical experts
  ☐ Yes
- Confirm the experts have travelled to the event source if possible
  ☐ Yes

RECTIFY PROBLEM — Hands-On Action

- Confirm all group members are taking action through experimentation and trial and error
  ☐ Yes
- Confirm the group applies the philosophy of fail fast to enable learning and further action
  ☐ Yes
- Confirm the group implemented a tracking system for tracking results
  ☐ Yes
- Confirm the group establishes a daily update on the status of the experimentation actions
  ☐ Yes
- Confirm the group establishes (at minimum) weekly updates with Senior Leadership on the status of the problem-solving actions
  ☐ Yes

This checklist is not intended to be exhaustive. Adaptions and modifications to fit local practice are encouraged.

Revision 1 / 2022

A. Situational Analysis – Frame the Problem. The first section calls for the independent Problem Tsar to be appointed, the communication process started, and the group of experts appointed.

B. Make Safe – Assemble a Group of Experts. The second section ensures the group of experts are clear about their role and responsibility and are empowered to take actions. A "fail fast" mindset is adopted and mistakes are expected.

C. Rectify Problem – Hands-On Action. The third section is designed to ensure action is taking place and the team is creating knowledge, performing second-order problem-solving, and conducting root cause analysis. Failure is framed as the lack of creating new knowledge through continuous learning.

Chapter conclusion and next steps

The T.R.U.S.T. solution is not a step-by-step process map, it is a set of tools that assumes (even encourages) mistakes to be made, but through the creation of trust, psychological safety, strong communication, and the fail-fast mindset, progress happens. The next chapter describes what happened when T.R.U.S.T. was used to solve a real S.H.I.T. problem in the field.

CHAPTER 6 T.R.U.S.T. in Practice

The test case was performed at Globomotive,[177] an automotive tier 1 supplier with a global presence. Globomotive is one of the 10 largest international automotive parts manufacturers in the world, producing a full selection of vehicle components. In 2019, in response to the electrification and digitisation of the automobile, Globomotive executed a tactical acquisition and partnership strategy to integrate autonomous driving electronics into its core business. The idea was grounded in the belief that cars will become autonomous and self-driving, and a competitive advantage should be pursued. The output of this strategy was the formation of a new business group, Globomotive Electronics.

A first realisation of this strategy was the design and manufacture of high-end automotive displays which integrate into the vehicle's instrument panel,[178] a standard Globomotive component. These high-end displays offer high-definition television quality and extend the diameter of the instrument panel. Design challenges include robustness of function during a vehicle's operating conditions, including weather and temperature extremes and driving terrain. Integration of the display design into the instrument panel design is critical, as the panel must be able to withstand both the display's weight and the heat it generates.

---

[177] Globomotive is a pseudonym to protect the anonymity of the company used in the test. Changes have been made to this case study to protect the anonymity of the people and businesses involved.

[178] An instrument panel can also be referred to as a car dashboard in the US or a fascia in U.K..

Figure 26 is an example of an instrument panel and the display in an autonomous driver vehicle.

Figure 26 Autonomous Vehicle Display Panel

In recognition of the Globomotive brand, and the complementary investments in mergers and acquisition adding new capability in design and manufacturing, Globomotive Electronics was able to secure its first display programs in Detroit, with 2019 volumes exceeding 300,000 displays. Due to the high-volume sales contracts and high ramp-up in Detroit, a non-automotive brownfield plant from a previous acquisition was secured for display industrialisation. The non-automotive plant had to be retrofitted to meet rigorous automotive requirements including staffing a new plant manager and management team as well as recruiting and training a production staff. In addition, manufacturing equipment had to be designed, sourced, and installed to meet demand.

Alpha Automotive Works is an American automotive company that produces a premium sedan broadly recognised throughout the world and associated with political leaders and elites. For its first major program out of the new plant in Detroit, the Globomotive displays team secured and contracted Alpha Automotive Works' newest luxury sedan. The development lead time for the new product was short. The formal milestone SOP (Start of Production) was contractually agreed for May 2020 with the car being made available to the public in August 2020. Priced between $97,700 and $133,100, the vehicle promised a new level of design indulgence, with the high-end display a primary marketing feature.

To summarise, the test scenario takes place in a large automotive corporation attempting to fulfil an order from a high-profile customer to design and manufacture new technology within a new business unit. The project was given high status and priority internally within Globomotive as well as within Alpha Automotive Works itself.

Genesis of a S.H.I.T. Problem - RED ALERT ISSUED

On March 18, 2020, at 13:17, a red alert escalation email was issued from the local Globomotive Program Quality Leader to the Globomotive electronics program team. The alarm was an alert communicating a severe event impacting the program: mass production manufacturing trials for the new displays had been completed and the result is 40% scrap; for every 10 pieces produced, 4 pieces were defective and placed in the waste bins. This result, with the program launch at Alpha Automotive Works less than 2 months away, put the program in jeopardy.

Events moved quickly as the problem became highly visible across the organisation; the following is a decription of select events reported by interviewees during the first 24 hours after the red alert was issued.

- 13:17 local time – RED ALERT issued – 40% scrap during the mass production trial runs.

- At 14:11, as the email distribution continued to expand and visibility increased, the Globomotive Program Manager issued a directive to the team to go faster and solve the problems ('Dear all, need quality team and engineering team support more and speed up the analysis and actions faster. Make it happen. Thanks.')

- Within 20 minutes (14:31) the Engineering Director responded that the problem was not his or Engineering's problem and someone else needed to take ownership and launch an 8D: 'Need to use 8D problem-solving methodology or Quality tools for analysis. Thanks.'

- At 19:11 the Program Quality Leader sent a personal note to the Globomotive leadership (Operations Director) explaining his situation: 'I'm like a fireman trying [to] put out fires everywhere handling customer requests for communication and update. Crazy! Everyone is criticising each other. Like battle!' Customer pressure was being applied to the Program Quality Leader. Alpha Automotive Works' Director of Purchasing became aware of the highly defective displays and demanded Globomotive senior leadership be aware of the risk and launch formal problem-solving 8D methodology. As was requested by the customer, the Globomotive Operations Director signed a document for Alpha Automotive Works acknowledging his recognition of the problem severity and need for 8D problem-solving.

- At 21:04 the first version of a very detailed technical 8D for the severe event was produced by the Globomotive program team and submitted to Alpha Automotive Works.

- At 23:02, after dozens of emails and escalation, the Globomotive Program Manager sent a recap to the leaders of Program Management (the Global Program Director) and Operations (Operations Director) for the division, that he pulled the team together that day for 'detailed issue solving' and that he personally was 'Tough and persistent, with finally success. 😊'. [Smiley face from the email.]

- At 23:26, the Globomotive U.S. Engineering Director forwarded a private email directly to the Division Management team that Engineering was not involved in the 8D problem-solving, and Engineering has rejected the report as this was not a design problem, and Engineering should not be blamed. He stated that, as the scrap was in the plant, the root cause must be plant and not design related, and Engineering would not take responsibility.

See the timeline below summarising the events during the first 24 hours:

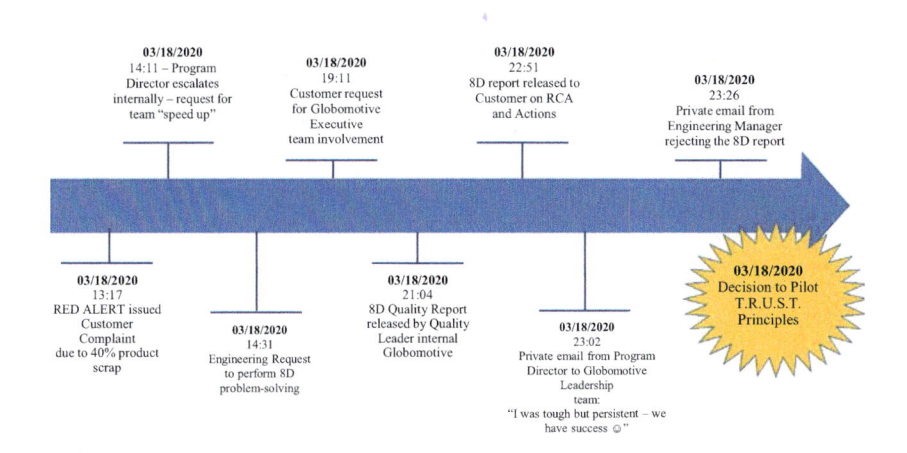

Figure 27 Timeline Day 1: Major events during the S.H.I.T. problem

8D problem-solving tried unsuccessfully

As was company procedure, the 8D problem-solving method was performed twice, unsuccessfully, by the Globomotive organisation. When interviewed as to why the 8D investigation was not successful the Globomotive Program Quality Leader responded:

> I tried to get the team all together, but no one wanted to touch this project. It was too high profile and dangerous. I had to create 8D on my own and send it to the customer. Everyone knows we had problems with a product defect, but people hide and I got stuck with the customer Alpha Automotive Works as I am Quality Leader and we have the day-to-day contact with the customer.

And when asked why he did not alert the management team before mass production trials that the display was having product issues, the Program

Quality Leader responded that 8D had already been tried before the complaint but the 8D was not successfully finding the root cause of the product scrap and pressure was building:

> '8D is my job. I do not want to get in trouble or fired because of a bad 8D. I did my best. It was not easy, so much pressure and stress with no support.' [Globomotive Quality leader]

As described below by the Globomotive plant manager, production of the defective displays continued even while the problems were known locally by the team because no one was taking ownership to drive and solve the technical issues:

> I remember we have lots of, how do you say, very worst state of quality .... [there was] at least 40% and sometimes 50% of scrap in a batch. We had many issues and faults in the product, in that time [March 2019] and we set up the 8D team to focus on the defect issues. .... To be honest we all knew we had the defect issues before March, and nobody was very clear about this and what was going on. Everybody thought that the defect issues were under analysis, but no extra actions had been taken. The 8D is just [a] report paper. It did not support us with fixing the product issues at all. [Globomotive plant manager]

The program team could not agree internally on the root cause analysis as multiple issues were present and the team began experiencing interpersonal conflicts and engaging in blame shifting. From the words of a Globomotive Senior Executive involved in the problem-solving:

> So, I think the team pretty much struggled to really assess the root causes of the quality issues we had in the ramp up phase of the program, meaning spots on the screen defects. And it took quite a while to acknowledge that there were several factors, not only our design. It was also linked to the supplier parts, the

> quality at the site, and to our manufacturing process. So suddenly, basically all functions had issues. Now, for example, initially it was very difficult to acknowledge for the engineering leader that his design may not be good enough. It was always another function and not engineering. So, there was this function ping, a finger pointed, and no accountability. [Globomotive Vice-President for Quality]

After several rounds of communication with the customer reviewing the Globomotive 8D, Alpha Automotive Works grew increasingly frustrated and on March 18th, after the Quality Alert was issued, the customer escalated the severity of the problem from their perspective to Globomotive executive committee and began to threaten and use aggressive behaviour towards the Globomotive program team (e.g., foul language). The conduct of the customer had a depreciatory effect, further adding pressure to the team. Escalation to the Globomotive executive committee composed of the top U.S. leadership resulted in the emergence of political forces as the different functions overtly avoided responsibility for the customer escalation and root cause of the display's technical issues. Manifold Microsoft PowerPoint decks were prepared for the executive committee explaining why their function was clean and operating according to process and procedures.

As a result of these customer escalations, the Globomotive display Operations Director partnered with the Vice-President for Quality and agreed that due to the severity of the problem and the need for a swift resolution, a different approach was needed. The Vice-President agreed to pilot the solution design: Solving S.H.I.T. Problems with Principles of T.R.U.S.T.

During an interview, the Vice-President made the following comments reinforcing the complexity and severity of the problem the team faced with the customer and the display product as the team was working on the edge of knowledge, managing a new product with new technology in a new plant:

> And probably we need to outline the bigger picture, which is that we were not only working on the time pressure for a new program …. it was [also] the first critical launch of a brand-new plant and…. the first time that that team was developing such a product. And it was the first time that this team worked with the technology that they were only familiar with through an acquisition that we did outside of the U.S., that initially transferred the technology to the new plant, but then was deployed back at the end of last year. And the Detroit team was more or less on its own for the first time. So, two things, new plant and team that is doing this for the first time. Then we also had the very, the plant itself had a very high turnover of plant managers. There was a plant manager that couldn't be in the plant [due to COVID travel restrictions], so we had an interim plant manager who also had the role of launch manager. [Globomotive Vice-President for Quality]

Additional external pressure and complexity: COVID-19

On November 17, 2019, the Chinese government officially reported the first case of COVID-19 coronavirus in the city of Wuhan, part of the Hubei province. By mid-January COVID cases were reported throughout China and by end of January the virus had spread to the U.S. and Europe.[179] The impact of COVID-19 was relayed from the Vice-President of Quality:

> And everything happened during the time of the COVID outbreak, which means initially the team had difficulties traveling. And later on, we couldn't get any global experts because no one could travel anymore [due to country COVID-19 travel bans]. ... The team was on its own. Also, politically the company went through a reorganisation at the same time, not only linked to COVID restructuring, but also simply by putting a display team into the regional divisions. This happens simultaneously leading to many...changes.... probably putting the people under even more severe personal pressure and feelings of, 'Will I have a job after this program? What will I do now? Will there be any jobs after COVID?' All of these elements added to the anxiety and nervousness...

S.H.I.T. problem conditions confirmed using the T.R.U.S.T. solutions

An assessment of the situation was done using the S.H.I.T. Problem-Solving Identification Evaluation Tool. The template was completed by the Operations Director for the Display Technologies division with input from the members of the functional team working on the program. The assessment confirmed the presence of S.H.I.T. conditions, which led to implementation of the rest of the solution for augmented problem-solving.

---

179 https://www.who.int/news-room/detail/27-04-2020-who-timeline---covid-19

168

Figure 28   S.H.I.T. Problem-Solving Identification Evaluation Tool completed

This evaluation should answer the question, "Is this a S.H.I.T problem for the business where traditional problem-solving needs augmentation?"

System Difficulty	Severe Event? Is the event severe to your business? Is the event impacting Health and Safety as well as Customer?		Highly Visible? Is the event visible in your internal company? Is the company in the news or other multi-media?		Information Deficient? Do you need to secure information? Can the interfaces be managed and controlled?		Time-Critical? Are there external or internal time constraints? Is the end-Customer impacted?	
	Minimum Criteria	Check	Minimum Criteria	Check	Minimum Criteria	Check	Minimum Criteria	Check
Routine	– Minimal Customer impact – Minimal Technical risk – Minor Financial exposure – Operations not impacted (no stop work required)		– Event is locally contained – Customer involvement not required – Internal Customer involvement / approval needed		– Problem repetitive and solution is readily available (no RCA needed) – No investigation needed - root cause known – Local expertise available		– Achievable in given time frame. – No actual internal or external customer impact affected	
Significant	– High probability of major technical issues – Significant repair work expected – Significant Financial exposure	X	– Customer escape - non-conformance left value stream – External Customer / authorized body approval needed for technical solution – Event highly visible in the organization		– Engineering interface required - root cause unknown – Multiple locations may be involved – Product or Process redesign needed		– External customer order affected, overall time schedule is jeopardized – Crisis recovery plan needed – Customer penalties a high-probability	
Complex	– Certainty for replacement of major components – Immediate stop required – Confirmed safety issue – Potential product recall - warranty event likely		– Company loss of reputation threatened - external media management needed – Customer involvement - 3rd party litigation high-probability – Serial production impacted / product recall	X	– Technical interfaces - cross functional team required – Multiple locations involved – Product redesign driving Production / Service process redesign (e.g. Design concept / technology redesign)	X	– Impact of external customer contractual dates jeopardized – Customer stop work enacted – Liquid damages / penalties actualized	X
Comment / Reason for selection	– Product scrap rate of ~approximately 40% – Stop required - product not viable for mass production with current technical issues		– Customer complaint written to the CEO of Globomotive – Globomotive board member overview on technical non-conformance – First launch in new plant – industry visibility to problem with media risk of Alpha Automotive Works not launch on-time		– Problem undefined with no common list with overview on technical non-conformance – No clear ownership of the problem internally in Globomotive		– Customer SOP (start of mass production) less than 60 days from red alert issued – Risk of Customer penalties to be incurred due to line stop if problem not solved	

As the completed template illustrates, all four conditions were confirmed and identified as the highest level of difficulty: complex. The comments in the bottom row show how real-world facts from the problem were characterised within the four categories. Users identified that the event was 'Severe' on the basis that the product scrap rate was approximately 40% and that an immediate production stop was required. The problem was considered 'Highly Visible' because a customer complaint was written to the CEO of Globomotive U.S., the launch in a new plant created high industry visibility and there was a risk of media coverage due to the high profile of the Alpha Automotive Works product. The problem was characterised as 'Information Deficient' on the basis that after two 8D efforts, no one had taken ownership of the problem and the cause of problem was unknown. Finally, the problem was considered 'Time-Sensitive' because the Customer SOP (start of mass production) was less than 60 days from the date the red alert was issued and there was a risk of customer penalties if the problem was not solved. Once the existence of a S.H.I.T. problem was confirmed, executive leadership employed the T.R.U.S.T problem-solving principles.

T.R.U.S.T. principles applied

To recap, the 8D problem-solving methodology was attempted twice, but failed to solve the technical issues with the display. Upon confirmation of S.H.I.T. conditions, executive leadership agreed to pilot the T.R.U.S.T. principles:

(T) Tsar appointed independently to govern and support the problem-solving team

(R) Re-frame failure as an opportunity for knowledge

(U) Understand and define the scope of the problem

(S) Support open communication structures

(T) Technical experts empowered to discover

The following sections narrate how the T.R.U.S.T. principles were employed to resolve the problem through the application of the designed solution. This chapter benchmarks progress implementing these principles using the T.R.U.S.T. In Action checklist.

The principles outlined in figure 29 were applied continuously and in parallel following the logic that the T.R.U.S.T. principles are not intended to be deployed only chronologically.

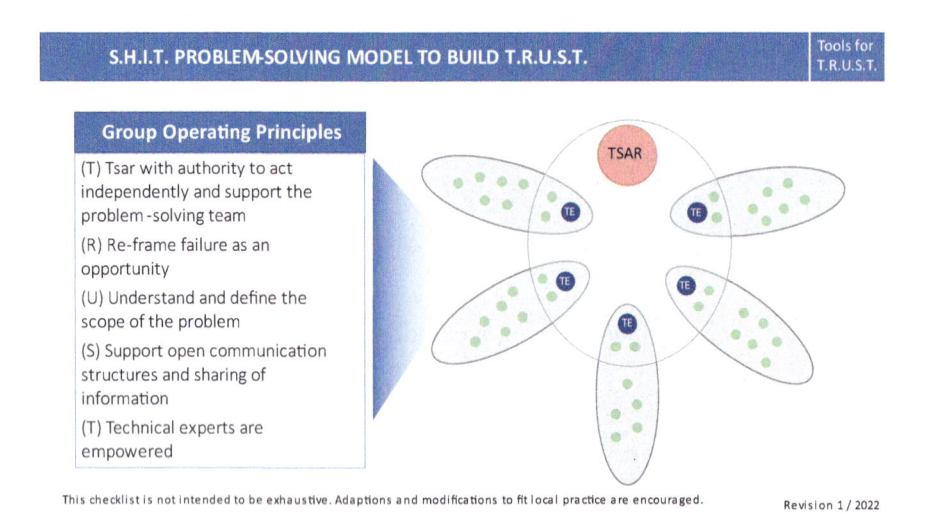

Figure 29 Model for solving S.H.I.T. problem using T.R.U.S.T. principles

(T) Tsar appointed independently to govern and support the problem-solving team

After the Customer complaint, the failure of multiple 8D root cause efforts, and the customer's subsequent escalation, an alternative approach was needed. The corporate Vice-President of Quality, also the former leader of the U.S. for Globomotive, was appointed by the CEO as the Problem Tsar.

The Problem Tsar recognised the sensitivity of the situation from an employee standpoint. Describing his initial support to the team, the Vice-President of Quality stated:

> Well, first of all we try to create a security net for them [the team], by making sure that they would not fail. For example, we always had a trial run before we went over the progress status with the ExCom [Executive Committee], and where we make sure the presenters would feel comfortable with the content. We also had a very smart way of not leaving everything to a program manager, but really involving each functional leader for each of the issues. So, every issue had the functional leader as pilot [action owner] that needed to report on that particular issue and the resolution plan, by doing so, the program manager was able to better engage the functional support and wider team and make it a priority for everyone to make this program a success. [Globomotive Vice-President for Quality]

The Program Manager, in a post-problem-solving interview, relayed several times how this focus on empowerment and inclusiveness by the executive leader impacted him personally and gave team members confidence and permission to speak openly and to be transparent.

After the customer's formal complaint and the feedback of the 40% scrap rate, the display technical issues became very visible in the company—beyond the display division. The Globomotive executive committee board feared that recognition in the marketplace of the potential failure of this new technology would bring the company bad external press and jeopardise the important U.S. market. This fear was accelerating due to the lack of real information on what the problem was. It was known that several 8Ds had been conducted with no impact on the product quality or non-conformance rates. In spite of these mounting pressures, the Problem Tsar maintained daily touchpoints with the problem-solving team and took action to protect the team, containing this pressure and keeping it separate from their efforts. Feedback from the team confirmed that having a Vice-President being formally appointed to support and coach the team reduced the amount of personal fear and pressure experienced by being close to the problem, as well as the perceived threat of political ramifications from being a member of the team. The plant manager emphasised the role of the Problem Tsar in providing protection and reducing fear:

> With [Problem Tsar] personally involved, we are less scared. This is important. Because scared cannot help us do anything. With fear, we're like, you protect yourself. Only, only, yourself. Without protection you will avoid, how do you say, 'this is not my responsibility, it's not my problem' … When you have this type of trouble, you need to focus on the issues, not the pressures and panic… [Globomotive plant manager]

When the Tsar was asked to reflect on which actions from the T.R.U.S.T. method were most impactful, he identified the environment of protection for the team and the effectiveness of a problem-solving team composed of colleagues from a variety of functions across the company.

Unexpected Outcomes

A key unexpected outcome of appointing an executive as Tsar of the problem-solving team was a shift in the willingness of people wanting to contribute to the problem-solving effort. Even though the problem was initially described as a 'time bomb' that could derail careers, once the organisation signalled the importance of the problem through the executive's appointment, organisational members were volunteering to join the problem-solving team. As the engineering director related, originally he felt very hesitant, even blocking the root cause analysis, but once the Problem Tsar was appointed, he assigned himself as the functional representative of engineering while also sending his management team and technical experts to the plant for the execution of design of experiments and product testing.

Milestones Completed from the T.R.U.S.T. In Action Checklist

1. Independent problem owner (i.e., Problem Tsar) appointed

2. Appointment by CEO provided Tsar with appropriate cross-functional authority

(T) A team of empowered technical experts is established

Cross-functional actors representing a team of teams are needed to understand and collectively apply the scientific knowledge of multiple organisational functions; in complex systems, specialisation is key. The

8D approach did not meet this need. In the words of the Vice-President of Quality:

> We clearly didn't make any customer progress with the 8D. And the issue with the 8D is that we would not have been able to holistically address the problems and to holistically improve multiple parameters that impact one system that works together that is completely interconnected. And probably that is becoming more and more the challenge as everything is a system. And this leadership cross-functional approach helped us to, to really make a difference and to go beyond traditional tools in automotive enabling the system to perform better. [Globomotive Vice-President for Quality]

As a first action, the Problem Tsar organised a 'team of teams' to work on the severe event. This included the director of operations, the plant manager where the parts would be produced, the engineering leader responsible for the product design, the program manager responsible for the product interface and the industrial manager responsible for the plant equipment. By involving all the functional leaders together as a core team, supporting experts were available immediately and as needed.

During the first meeting with the team, the Vice-President of Quality explicitly promoted psychological safety by acknowledging that he was not an expert in the product or technology and clarifying that his role was to support the team by removing roadblocks as they came. The team was fully empowered to make and take decisions and the Problem Tsar relied on their technical expertise to address the technical issues in the problem. The Problem Tsar officially granted the team permission for openness and transparency when working together. In addition, as described in the T.R.U.S.T. in Action Checklist, active team building exercises were employed so that each member was known by name, role, expertise, and experience.

In order to formally communicate 'what the problem is' to the organisation, the Problem Tsar organised the team to prepare a situational analysis containing the critical action list (in PowerPoint) for distribution and communication within the company. This informed all stakeholders on the progress and signalled explicitly to the wider organisation the team was moving forward. Each technical problem was summarised in one slide, with a picture and non-technical jargon for easy communication. To give confidence to the stakeholders, the efforts of the team were made visible, including development of a list of current open points and owners of those topics. In addition, an experiment was started to isolate the problem in order to create knowledge through action. This experiment was added to the PowerPoint and the status was updated every week to communicate what had been learned and how that learning made a direct contribution to a reduction in scrap.

As the team was not used to working together, this first act of creating a common viewpoint among the functions required many hours of preparation and revision (estimated at approximately 40 hours of preparation by the Tsar and each team member). The views and insights of team members were continually invited, which created alignment and facilitated teamwork.

In a decisive act of sensegiving, the Problem Tsar organised a bi-weekly product readiness review which was extended to the entire global organisation of potentially impacted stakeholders by Alpha Automotive

Works. This included teams in business units in the U.K., France, Germany, Taiwan, China, Japan, and, of course, the U.S. This helped ensure a consistent message across all parties and stakeholders and had a secondary effect of demonstrating organisational support of, and investment in, the success of the problem-solving team. Demonstrating the active role of the team leader, delivering psychological safety through inclusive leadership, investing time in coaching, and team building were all strategies employed by the Problem Tsar. For example, the Problem Tsar coached the team by preparing them for board-level executive committee progress reviews.

The team of technical experts each had a role to play in the problem-solving process. The plant manager explained how valuable it was to bring together these multiple organisational functions:

> Even if you are the very professional guy who knows everything, if you cannot get the support from the other team, you cannot solve these issues very quickly. For example, you know, you need the data set, and you need the operators and the plant engineer to help you to implement your actions, day by day. You cannot do everything by yourself, for sure. Not in a crisis. [Globomotive plant manager]

The plant manager observed how the cross-functional 'team of teams' approach facilitates the problem-solving process:

> The reason we must use teamwork is because maybe the plant team cannot provide some technical solutions. But they could support to do the measurement. Support other functions to take some actions. That is very important to make progress.... That's

> the difference I have seen using this approach [T.R.U.S.T.].
> [Globomotive plant manager]

Unexpected Outcomes

A significant unexpected outcome was that when presented with the T.R.U.S.T. problem-solving approach, the team readily and eagerly gave up the traditional approach to 8D for problem-solving the technical issues. When asked during the post-mortem interviews why, after so many years of telling people they must do 8D, the traditional problem-solving process was so readily abandoned, the plant manager explained:

> 8D is just a final reports paper to have our thinking show [the] customer we solved the issues. But when we do not know the cause, how can we solve the issue with 8D reports? How can it help us? And I think the best way [to solve problems] is to deeply involve the team in the solutions. And that's the key, how can we solve the issues, not the reports. An 8D report is just one paper. ... 8D reports cannot help us in the beginning of that issue, a team is needed because nobody knows how we can solve the issues. This is what we did, the team solved the issues.
> [Globomotive plant manager]

The plant manager describes 8D as a tool with limited usefulness that does not solve a problem when the cause is unknown. The T.R.U.S.T. approach purposefully brings together a team of people with the necessary expertise to solve a problem and creates an environment where they can effectively collaborate. Further, as the interviews demonstrated, the 8D approach of gathering information for a report is process-driven, rather than outcome driven. The report was refuted, and self-preserving behaviours were triggered: the process itself generated stress for the people involved, potentially delaying resolution of the problem and negatively impacting employees and the corporation.

Discussion of Milestones Completed from the T.R.U.S.T. In Action Checklist

1. Elements of the permission charter were completed, and the technical experts were empowered. The Problem Tsar applied the principles of inclusive leadership by acknowledging personal limitations of technical knowledge and creating an inclusive team environment where all members are invited to contribute regardless of job level.

2. A problem-solving team was established from the heads of each functional organisation; a 'team of teams'.

3. Problem-solving was treated as a cross-functional, collaborative project. The project manager had the support of a Problem Tsar who functioned as a coach and ally.

4. A Severe Event alert was communicated. With the Problem Tsar's support, the cross-functional team prepared a full situational analysis of the problem and created and formally documented it into a critical issues list in PowerPoint. Each issue was summarised with a PowerPoint slide with pictures, verbal descriptions, and action owners.

(S) Support open, streamlined communication structures

Sharing of real-time information about the context of the problem and the actions taken builds trust and creates opportunity for critical challenge which can help avoid 'group think' and other biases. The T.R.U.S.T principle of open and streamlined communication structures was advanced with daily cross-functional 'top 5' updates and the utilisation of social media applications like WhatsApp, which allow for instant sharing of information.

In a crisis, effective communication becomes extremely important, and by reinforcing messages with multiple communications, teams can help to unify the entire organisation around the handling of the problem. In this case, the CEO himself engaged in daily reviews, supporting the team with meaningful questions and attention.

Consistent with the research findings and 3P framework, S.H.I.T. conditions initially divided the Globomotive teams and people responded with blame shifting and an unwillingness to share information (a cost of 3P). In fact, right after the red alert was issued, the Project Quality Leader privately informed the Problem Tsar that the engineering group had not been truthful with the customer. While communicating this news, the Project Quality Leader also attempted to recuse himself from involvement in the problem and discharge his responsibility.

To mitigate these responses and promote open and transparent communication, the Problem Tsar encouraged the teams from Engineering to temporarily move from their central offices and design centre in Chicago to the plant in Detroit. This approach facilitated immediate feedback on actions taken during the trial-and-error process and improved trust and cohesiveness between team members. During this time it was the height of the COVID-19 pandemic in the U.S. This decision to allow the team to travel was only possible because a Tsar had been empowered by the executive board of the company. Further, the Tsar personally travelled to the plant regularly to demonstrate inclusive leadership by working alongside the team and sharing the risk of exposure to COVID-19.

The importance of communication during a severe event cannot be overstated. The Alpha Automotive Works event affected the company's internal functions (i.e., Engineering, Sales, Supply Chain, Production, Quality, Marketing, Programs), multiple business segments within the company; various geographic regions of the company (i.e., Japan, the U.S., Europe) and external stakeholders such as the customer. As the research illustrated, a problem with broad-reaching impacts affecting a wide range of functions and business segments could have resulted in teams acting at cross purposes and falling back on 3P responses. The company obviated this outcome by employing the T.R.U.S.T. principles to support the sharing of information and open communication. A strong multi-channel approach to communications (further described in the milestones below) was implemented to temper escalations and interjections into the root cause analysis. The team was given direct access to the CEO, who engaged in daily reviews with the team, supporting them through meaningful engagement and attention and minimising opportunities for misinformation to take hold.

A severe event explodes awareness among various interests and as sensemaking tells us, 'perception is reality' and can derail the root-cause analysis. The company obviated this effect by taking a strong multi-channel approach to communications as various interests tried to escalate and interject into the root cause analysis.

As time progressed and the team gained confidence through the protection and encouragement of the Problem Tsar, cohesiveness increased and reinforced the empowerment of the team of technical experts. At the beginning of the Red Alert, the Operations Director sent a daily recap in email on the highlights of the day. This key activity of sensemaking was soon assumed by the Program Manager who, through his daily meetings with the team, created an action list with owners and dates with action severity, which he distributed daily to the severe event stakeholders.

Over time the team took on some of the roles initially filled by the Tsar, however the environment of psychological safety that was developed early in the problem-solving endeavour persisted, which helped sustain the team's confidence and the expectation that they would receive support. The Plant Manager commented on the value of streamlined communications:

> The top management, they cannot focus on all these details. They cannot have a meeting with us every day because they have a lot of meetings to attend. What they did here was encourage and motivate. They support us when we need it. We know they are there for us when we need it. We are all in this together... [Globomotive plant manager]

Unexpected Outcomes

One unexpected outcome with positive results was the usefulness and power of social media, versus traditional business communication such as email, to drive communication that was more informal, more honest, and in real time. The WhatsApp group was created with all functions and stakeholders in the very beginning of the discovery of the severe event, right after the red alert was issued. As updates were discovered during the day, they were instantly communicated to all. And as the stakeholders grew, so did the list of members in the WhatsApp group. Finally, whenever the team needed an ad-hoc meeting, instead of opening Outlook in a laptop or going to an office, they used the social media app to conduct impromptu meetings, including meetings with video conferencing. PowerPoint decks were replaced with video and voice. Finally, as the Problem Tsar and the Operations Director were part of the social media group, they were able to make and take comments from the team—and did. The informal communication on WhatsApp removed barriers and heightened psychological safety.

Milestones Completed from the T.R.U.S.T. In Action Checklist

1. Rapid communication tools created. Two WhatsApp groups were established. One with the customer for immediate feedback and the second with the problem-solving team and stakeholders.

2. Daily update on the status of the experimentation actions. Daily morning expert technical reviews were held with the cross-functional team to review the status of the investigation and learnings.

3. Information withholding and distortion were avoided. A daily evening review was held for the core team and invited stakeholders tracking the action list.

4. Daily evening report-out email. Wins and challenges for the day were shared and the next day's actions identified.

5. Weekly update for Executive Leadership. A formal, weekly meeting with the CEO and the leadership board on the status of the problem-solving was convened. Key actions and needed support were also addressed.

6. Travel to the event source. The Problem Tsar travelled to the plant approximately every two weeks to see the team face to face.

7. Travel to the event source. The Problem Tsar approved the travel expense and all of the technical team members were allowed to temporarily relocate to the plant.

(R) Re-frame failure as an opportunity for knowledge

With a S.H.I.T. problem there is inadequate information to determine the root cause. The T.R.U.S.T. principle of reframing failure as a learning opportunity promotes psychological safety and encourages experimentation and creative thinking. Following the principles of sensemaking, actions create knowledge: knowledge of what failed and must be improved. Every failed trial and error yields new understanding and information, knowledge that brings the team closer to understanding the root cause of the problem. By reframing failures as knowledge and opportunity, the company makes it safe to share this new information widely among the team and with others who can help the company more quickly discover a solution.

The Globomotive Tsar reinforced this principle by supporting and praising experiments, even those that did not lead to issue resolution. When dealing with high-stakes, complex problems, every lesson learnt can help avoid significant consequences and bring the team closer to a solution.

The Globomotive team applied the lessons from each failed attempt, turning 'brainstorming' into 'trystorming' to further understand the scope of the problem. The plant manager explained how this iterative approach of trying, failing, and learning incrementally led to solutions:

> We can see progress on these issues daily. We can update our actions or optimise our actions every day with[in] the results. We solved the issues. Actually, we think this approach [the T.R.U.S.T. approach] is the new standard. These solutions and tools will be the standard for all [of] the more important issues. It works. [Globomotive plant manager]

As the technical team of experts continued to learn and problem-solve, the product quality continued to improve towards customer expectations. By highlighting this positive trend and emphasising ongoing actions, the team was able to communicate beneficial progress to both the leadership and the customer. In addition, in the weekly reviews with senior leadership, the business CEO supported reframing failure through positively reinforcing the team's efforts by congratulating them on each step of the process.

Week after week the team continued hitting the short-term targets and, rather than being in a pressure-cooker environment they instead congratulated each other and shared the progress with the entire organisation, enhanced by the social media WhatsApp group where videos and pictures were real-time. On May 5th, 25 days before mass production of the automobile, the team shared openly the status that they are not fully product-launch ready, but had actions defined to continue their progress. This message was shared with the customer, and with all the internal functions.

The effectiveness of the T.R.U.S.T. approach was evident externally as well. Within a month of augmenting the unsuccessful 8D problem-solving effort with the T.R.U.S.T. principles, Alpha Automotive Works, the external customer, recognised a sea change in the direction of the project and formally cancelled the red alert.

The evidence of forward progress within the firm and the customer's positive response reinforced the creation of psychological safety, helping to overcome the deleterious 3P effects commonly associated with S.H.I.T. problems.

Unexpected Outcome

One unexpected outcome of removing obstacles typically present in the 8D process was that the process of resolving the problem progressed faster than is typical when multiple defects are discovered. Re-framing failure as opportunity and facilitating cooperative group thinking eliminated many of the 3P costs normally associated with a S.H.I.T. problem. Instead of multiple different people acting divisively in their own self-interest, the new approach encouraged people to act constructively toward a common goal. The plant manager stated:

> For sure, I'm very proud of it. We started the actions to solve the scrap defect issues from April and then we have several products and for every product it would take around one month to solve each of these issues. What happened normally, with other

problems like this, is it may take more than a year to solve all of these problems, but after we set up the team and we focused on these issues it actually took one month, within one month, to solve very difficult technical issues. And it's proved that for sure this new way [T.R.U.S.T. principles] works. I'm very proud of that. This problem was difficult, with multiple causes, and we did it. [Globomotive plant manager]

Milestones Completed from the T.R.U.S.T. In Action Checklist

1. Reframing the problem. The Problem Tsar communicated within the team as well as to stakeholders, framing the problem as an opportunity to show U.S. capabilities internationally.

2. Traveling to the event source. Even with a crisis like COVID unfolding where international travel and support was limited, by solving the problem locally, the team was able to demonstrate their agility and robustness to disruptions.

3. Fail fast to enable learning. The group established a daily update on the status of experiments, emphasising learning as positive progress.

(U) Understand and define the scope of the Problem

The time pressure and lack of information that accompany a S.H.I.T. problem can give rise to paralysing fear and prevent the problem-solving team from acting. The problem-solving team is able to overcome the paralysing effect of inadequate information by augmenting the traditional problem-solving process with the T.R.U.S.T. principle of understanding and defining the scope of the problem. Applying this approach, the team was able to act despite having insufficient information. This action was focused on building understanding through trial and error combined with a structured approach for tracking new learning, communicating information, and reporting progress. The team pursued understanding by first articulating an imperfect problem statement based on the problem symptoms experienced (e.g., scrapped product) in an effort to begin to better understand the problem. While the earlier 8D approaches had started with a problem-statement and transitioned immediately to a root cause analysis, the Alpha Automotive Works investigation drove a structured trial-and-error design of an experiment with rapid failures and quick lessons to fully understand the scope and the characteristics of the problem, versus just the problem symptoms and the end result.

The approach taken as relayed by the Tsar:

> The only way to really measure progress is to define your design of [the] experiment in a very structured way where you document every part of the parameters you change. If you don't do that, you will never find the answer [with these types of complicated problems]. And it took some time to have this structure, which could only be realised through very good leadership, very strong leadership. [Globomotive Vice-President for Quality]

The Tsar reflected on the effectiveness of the 'fail-fast' approach to better understand the problem:

> If you have many factors that impact the quality and performance, if you need to have such a big design of [the] experiment, the only way to quickly get to results is to basically go quickly through experiments using trial and error to rapidly exclude certain dimensions, certain parameters that may not lead to anticipated results. With customer escalation and pressure

> from our very top management, the team was able to find the right process parameters, address the root causes in multiple functions simultaneously, and ramp up a successful program. [Globomotive Vice-President for Quality]

The cross-functional expert team furthered the T.R.U.S.T. principle of understanding and defining the scope of the problem by analysing the defects on the previous products and building a structured visual diagram highlighting important factors. These factors were then selected to be varied by trial and error and then the outcomes were measured for success.

Next, the team built a spreadsheet for tracking these various factors and the results of each test. Each of these actions generates a cue for further actions and making sense of the problem. This cycle of action, test, learn, and repeat moved the investigation forward. After many rounds of improvement, the design process was then codified and standardised for operator training. This approach led to progress, week after week: from 40% scrap to > 98% yield.

After a customer red alert and application of two successive 8D problem-solving efforts, the company was able to apply the principles of T.R.U.S.T. and meet the customer requirements of mass production by the May 2020 SOP. In less than two months the team was able to deliver a rapid problem-solving initiative and, in the process, establish new baselines of learning and growth. As could be expected, the congratulations from the various parties kept pouring in as the team was able to deliver what was thought impossible.

The Problem Tsar was so pleased with the results, he recommended standardising the application of T.R.U.S.T. as a common tool when dealing with these types of problems. He specifically identified the cross-functional focus and fail-fast design of the experiment approaches as key to rapidly resolving technical problems. In his words: '...the cross-functional design of experiment has been the critical turnaround for us.' [Globomotive Vice-President for Quality]

Unexpected Outcomes

An unexpected outcome was that the full technical scope of the problem was not fully understood by the team until all trial-and-error experiments were complete and the problem was resolved. Application of the T.R.U.S.T. principles led to a successful resolution of the problem even as the team continued to modify and update the problem statement in response to new information created during the cycle of learning.

Milestones Completed from the T.R.U.S.T. In Action Checklist

1. A fail-fast philosophy was adopted. The problem-solving team began by defining the problem during the root cause analysis problem-solving in order to create common understanding.

2. Action was taken through trial and error. The team employed a structured trial-and-error approach, measuring the cause and effect of each design and process change.

3. A situational analysis was completed. A full 5W2H situational analysis was performed and released on June 1st, the day the product was formally released to the customer (the start of production date).

Usefulness of the T.R.U.S.T. In Action Checklist

Throughout the problem-solving efforts, the Problem Tsar and the Operations Director referred to the T.R.U.S.T. In Action Checklist and implemented every recommendation on the list. The checklist proved effective for guiding the team and a tool for maintaining accountability and adherence to the T.R.U.S.T. principles.

191

Figure 30 Checklist developed for S.H.I.T. problems

T.R.U.S.T. IN ACTION CHECKLIST

Tools for T.R.U.S.T.

SITUATIONAL ANALYSIS

Create the Problem

- ✓ Confirm Tsar has cross-functional authority
  ☑ Yes
- ✓ Confirm Tsar completes 5W2H Situational Analysis
  ☑ Yes
- ✓ Confirm Tsar officially communicates findings from S.H.I.T. Problem-Solving Identification Evaluation to relevant senior personnel
  ☑ Yes
- ✓ Confirm Severe Event alert distributed to all relevant internal stakeholders
  ☑ Yes
- ✓ Confirm cross-functional group of experts appointed to team
  ☑ Yes
- ✓ Confirm rapid communication tools created (e.g. messaging group, daily meetings...)
  ☑ Yes

MAKE SAFE

Assemble Group of Experts

- ✓ Confirm all group members have introduced themselves by name and role – cross-functional team of teams from relevant functional areas
  ☑ Yes
- ✓ Confirm Tsar completes permission charter to promote a safe environment and verbally communicates permission to speak openly
  ☑ Yes
- ✓ Confirm Tsar reframes failure as opportunity for knowledge and promotes continuous learning through trial and error
  ☑ Yes
- ✓ Confirm decisions lie with technical experts
  ☑ Yes
- ✓ Confirm the experts have travelled to the event source if possible
  ☑ Yes

RECTIFY PROBLEM

Hands-On Action

- ✓ Confirm all group members are taking action through experimentation and trial and error
  ☑ Yes
- ✓ Confirm the group applies the philosophy of fail fast to enable learning and further action
  ☑ Yes
- ✓ Confirm the group implemented a tracking system for tracking results
  ☑ Yes
- ✓ Confirm the group establishes a daily update on the status of the experimentation actions
  ☑ Yes
- ✓ Confirm the group establishes (at minimum) weekly updates with Senior Leadership on the status of the problem-solving actions
  ☑ Yes

This checklist is not intended to be exhaustive. Adaptions and modifications to fit local practice are encouraged.

Revision 1/2022

Application of Six Sigma tools for process optimisation after mass production begins

Learning by trial and error was effective in problem-solving the technical root cause, however once production had stabilised and the May start-of-production deadline was met, the local team organised a Six Sigma Black Belt project for additional improvement to drive manufacturing process efficiency and cost reduction. This suggests the possible co-existence of S.H.I.T. problem-solving and Six Sigma.

Continued use of T.R.U.S.T. principles for an expanded scope of technical problems

During the post-problem-solving interviews, the team communicated that the traditional 8D approach had not been effective in team-based problem-solving and that going forward they were applying the principles of T.R.U.S.T. in the context of less severe problems. This suggests that there is potential for additional applications of the solution beyond severe events and S.H.I.T. problems and that further study and analyses is warranted.

Sustained production performance indicates the T.R.U.S.T. solutions resolved the problems

The longer-term analysis of the product revealed a standard production quality of over 99% to customer requirements. The technical solutions implemented resolved the problem.

Application of the T.R.U.S.T. principles facilitates open and transparent communication, which can identify customer issues earlier, deter finger pointing and blame shifting, and facilitate open dialogue, all of which can lead to quicker discovery of the root cause of the problem. This approach to communication is a key tool for mitigating high-pressure conditions and helping to ensure equal access to information, aligning the organisation around productive problem-solving and revealing the facts that will be most helpful for identifying the root cause. We summarise these points in the next and final chapter.

CHAPTER 7 CONCLUSION

This chapter marks the end of this book, but the beginning of something more. You've been introduced to ideas and now you must decide what to do with that information. How will you act on what you've learned? How will you turn knowledge into skills that can be applied?

Like anything, this can be done with practice.

Just like the T.R.U.S.T. approach to problem-solving, this transition from knowledge to practice should be viewed as an iterative, continuous learning effort. Implementation of these ideas is not a win-or-lose proposition, but an exercise of improvement over time. I've been testing and using the T.R.U.S.T. approach for several years now and have learned some lessons along the way that may help you.

There are many ways to get started, but I think one of the easiest is to begin with S.H.I.T. Problem-Solving Identification Evaluation. Start looking at problems with this lens so that you can develop the skills to identify when a problem you are facing might require an augmented approach.

Another early step is to start expanding the problem-solving team to draw on the knowledge of technical experts. Increased communication is another step in the T.R.U.S.T. process that's likely to be very practical. If you are a leader of problem-solving efforts, consider how you frame the problem and the act of problem-solving itself. You likely have the autonomy to implement many or all of the leadership techniques of a T.R.U.S.T. Tsar.

While most of the elements of T.R.U.S.T. will fall within the authority of a quality professional to adapt at-will, some will be more challenging. In my personal experience, protecting the problem-solving team from outside pressure and creating a safe space has sometimes been difficult. There can be tremendous pressure from senior leaders to proceed and to resolve these high-stakes problems; standing in front of that barrelling train can be daunting. There are techniques built into T.R.U.S.T. that can alleviate the pressure from others to decide and act quickly. I have found that performing tests and communicating frequently shows both action and improvement toward an outcome, which can help give people confidence that things are moving in the right direction. But there will be times when, even using these tools, you will face tremendous pressure to act, and/or to make a decision when you don't have enough information to do so. For this challenge, I offer the following advice: be brave. Courage under pressure is often rewarded and is necessary for effective leadership. You can do this. To find support in this, and other steps in implementing the T.R.U.S.T. approach, there is an online professional network of practitioners. This network includes quality professionals from around the world, who look to find new ways to tackle big problems. Here you can ask questions, share stories, and meet others in the same boat. I also offer a periodic newsletter featuring case studies, interviews, stories, and updates from people working the T.R.U.S.T. approach in the field.

In addition to the efforts of individual practitioners, companies can adopt and adapt the S.H.I.T. problem framework and the T.R.U.S.T. approach and tools. Training, support, and guidance is available to companies seeking to better position themselves to face today's complex crises. Sign up for the newsletter and find more information at theshitshow.biz.

Good luck!

James Ezhaya

BIBLIOGRAPHY

Argyris, C. (1977), 'Double Loop Learning in Organizations', *Harvard Business Review*, 55 (5): 115-125.

Ariely, D. (2008), *Predictably irrational* (New York: HarperCollins)

Barthol, R. P., and Ku, N. D. (1959), 'Regression under Stress to First Learned Behaviour', *Journal of Abnormal and Social Psychology*, 59 (1): 134-136.

Bartz-Kurycki, M. A., Anderson, K. T., Abraham, J. E., Masada, K. M., Wang, J. S., Kawaguchi, A. L., Lally, K. P., and Tsao, K. (2017), 'Debriefing: the forgotten phase of the surgical safety checklist', *Journal of Surgical Research*, 213, 222-227.

Bearman, M., (2019), 'Focus on Methodology: Eliciting rich data: A practical approach to writing semi-structured interview schedules', *Focus on Health Professional Education – A Multidisciplinary Journal*, 20 (3): 1-11.

Beck, T. E., and Plowman, D. A. (2009), 'Experiencing Rare and Unusual Events Richly: The Role of Middle Managers in Animating and Guiding Organizational Interpretation', *Organization Science*, 20 (5): 909-924.

Berthod, O., and Muller-Seitz, G. (2018), 'Making Sense in Pitch Darkness: An Exploration of the Sociomateriality of Sensemaking in Crises', *Journal of Management Inquiry*, 27 (1): 52-68.

Best, M., and Neuhauser, D. (2006), 'Walter A Shewhart, 1924, and the Hawthorne factory', *Quality & safety in Healthcare*, 15 (2): 142-143.

Blatt, R., Christianson, M. K., Sutcliffe, K. M., and Rosenthal, M. M. (2006), 'A sensemaking lens on reliability', *Journal of Organizational Behavior*, 27 (7): 897-917.

Blumenthal-Barby, J. S. (2016), 'Biases and Heuristics in Decision Making and Their Impact on Autonom',. *American Journal of Bioethics*,

16 (5): 5-15.

Blumenthal-Barby, J. S., and Krieger, H. (2015), 'Cognitive Biases and Heuristics in Medical Decision Making: A Critical Review Using a Systematic Search Strategy', *Medical Decision Making*, 35 (4): 539-557.

Bowerman, E. R., and Littauer, S. B. (1956), 'Operations Engineering', *Management Science*, 2 (4): 287-298.

Boyer, K. K., and Pronovost, P. (2010), 'What medicine can teach operations: What operations can teach medicine', *Journal of Operations Management*, 28 (5): 367-371.

Brinkmann, S. (2013), 'The SAGE Handbook of Interview Research: The Complexity of the Craft', *Acta Sociologica*, 56 (4): 375-376.

Bryman, A., Foster, L., Sloan, L., and Clark, T. (2016), *Bryman's Social Research Method,s 6e*.

Bucknall, T. K., Forbes, H., Phillips, N. M., Hewitt, N. A., Cooper, S., Bogossian, F., and Investigators, F. A. (2016), 'An analysis of nursing students' decision-making in teams during simulations of acute patient deterioration', *Journal of Advanced Nursing*, 72 (10): 2482-2494.

Bui, T. C., Krieger, H. A., and Blumenthal-Barby, J. S. (2015), 'Framing effects on Physicians' Judgment and Decision Making', *Psychological Reports*, 117 (2): 508-522.

Cabral, R. A., Eggenberger, T., Keller, K., Gallison, B. S., and Newman, D. (2016), 'Use of a Surgical Safety Checklist to Improve Team Communication', *Aorn Journal*, 104 (3): 206-216.

Cannon, M. D., and Edmondson, A. C. (2001), 'Confronting failure: antecedents and consequences of shared beliefs about failure in organizational work groups', *Journal of Organizational Behavior*, 22 161-177.

Cannon, M. D., and Edmondson, A. C. (2005), 'Failing to learn and learning to fail (Intelligently): How great organizations put failure to work to innovate and improve', *Long Range Planning*, 38 (3): 299-319.

Cannon-Bowers, J. A., and Salas, E. (2001), 'Reflections on shared cognition', *Journal of Organizational Behavior*, 22 195-202.

Carmeli, A., Tishler, A., and Edmondson, A. C. (2012), 'CEO relational leadership and strategic decision quality in top management teams: The role of team trust and learning from failure', *Strategic Organization*, 10 (1): 31-54.

Carroll, J. S., and Edmondson, A. C. (2002), 'Leading organisational learning in health care', *Quality & Safety in Health Care*, 11 (1): 51-56.

Carter, N. M. (1986), 'General and Industrial Management, Henri Fayol', *The Academy of Management Review*, 11 (2): 454-456.

Casey, M., Leary, D., and Coghlan, D. (2018), 'Unpacking action research and implementation science: Implications for nursing', *Journal of Advanced Nursing*, 74 (5): 1051-1058.

Chia, R. (2000), 'Discourse analysis as organizational analysis', *Organization*, 7 (3): 513-518.

Christianson, M. K. (2019), 'More and Less Effective Updating: The Role of Trajectory Management in Making Sense Again', *Administrative Science Quarterly*, 64 (1): 45-86.

Christianson, M. K., and Barton, M. A. (2021), 'Sensemaking in the Time of COVID-19', *Journal of Management Studies*, 58 (2): 572-576.

Christianson, M. K., Farkas, M. T., Sutcliffe, K. M., and Weick, K. E. (2009), 'Learning Through Rare Events: Significant Interruptions at the Baltimore & Ohio Railroad Museum', *Organization Science*, 20 (5): 846-860.

Christianson, M. K., Sutcliffe, K. M., Miller, M. A., and Iwashyna, T. J. (2011), 'Becoming a high reliability organization', *Critical Care*, 15 (6): 5.

Clarke, L., and Perrow, C. (1996), 'Prosaic organizational failure', *American Behavioral Scientist*, 39 (8): 1040-1056.

Collatto, D. C., Dresch, A., Lacerda, D. P., and Bentz, I. G. (2018), 'Is

Action Design Research Indeed Necessary? Analysis and Synergies Between Action Research and Design Science Research', *Systemic Practice and Action Research*, 31 (3): 239-267.

Collins, K. M. T., Onwuegbuzie, A. J., and Jiao, Q. G. (2007), 'A Mixed Methods Investigation of Mixed Methods Sampling Designs in Social and Health Science Research', *Journal of Mixed Methods Research*, 1 (3): 267-294.

Conley, D. M., Singer, S. J., Edmondson, L., Berry, W. R., and Gawande, A. A. (2011), 'Effective Surgical Safety Checklist Implementation', *Journal of the American College of Surgeons*, 212 (5): 873-879.

Corley, K. G., and Gioia, D. A. (2011), 'Building Theory about Theory Building: What Constitutes a Theoretical Contribution?', *Academy of Management Review*, 36 (1): 12-32.

Costa, E., Soares, A. L., and de Sousa, J. P. (2018), 'Exploring the CIMO-Logic in the Design of Collaborative Networks Mediated by Digital Platforms', *Collaborative Networks of Cognitive Systems*, 534, 266-277.

Croskerry, P. (2003), 'The importance of cognitive errors in diagnosis and strategies to minimize them', *Academic Medicine*, 78 (8): 775-780.

Croskerry, P. (2013), 'From Mindless to Mindful Practice - Cognitive Bias and Clinical Decision Making', *New England Journal of Medicine*, 368 (26): 2445-2448.

de Mast, J., and Lokkerbol, J. (2012), 'An analysis of the Six Sigma DMAIC method from the perspective of problem solving', *International Journal of Production Economics*, 139 (2): 604-614.

de Vries, E. N., Prins, H. A., Crolla, R., den Outer, A. J., van Andel, G., van Helden, S. H., Schlack, W. S., van Putten, M. A., Gouma, D. J., Dijkgraaf, M. G. W., Smorenburg, S. M., Boermeester, M. A., and Grp, S. C. (2010), 'Effect of a Comprehensive Surgical Safety System on Patient Outcomes', *New England Journal of Medicine*, 363 (20): 1928-1937.

de Vries, E. N., Ramrattan, M. A., Smorenburg, S. M., Gouma, D. J., and Boermeester, M. A. (2008), 'The incidence and nature of in-hospital adverse events: a systematic review', *Quality & Safety in Health Care*, 17 (3): 216-223.

Deming, W. E. (1975a), 'Probability as a basis for Action', *American Statistician*, 29 (4): 146-152.

Deming, W. E. (1975b), 'Some Statistical Aids toward Economic Production', *Interfaces*, 5 (4): 1-15.

Denyer, D., Tranfield, D., and van Aken, J. E. (2008), 'Developing design propositions through research synthesis', *Organization Studies*, 29 (3): 393-413.

Detert, J. R., and Edmondson, A. C. (2007), 'Organizational behavior - Why employees are afraid to speak', *Harvard Business Review*, 85 (5): 23.

Detert, J. R., and Edmondson, A. C. (2011), 'Implicit Voice Theories: Taken-For-Granted Rules of Self-Censorship at Work', *Academy of Management Journal*, 54 (3): 461-488.

Dobrin, C., Girneata, A., Mascu, M., and Croitoru, O. (2015), 'Quality: A Determinant Factor of Competitiveness – the Evolution of ISO Certifications for Management Systems'

Donaldson, M. S., Corrigan, J. M., and Kohn, L. T. (2000), *To err is human: building a safer health system* (National Academies Press).

Dunne, D. D., and Dougherty, D. (2016), 'Abductive Reasoning: How Innovators Navigate in the Labyrinth of Complex Product Innovation', *Organization Studies*, 37 (2): 131-159.

Durivage, M. A. (2016), *Practical Design of Experiments (DOE): A Guide for Optimizing Designs and Processes* (Quality Press).

Dutton, J. E., and Ashford, S. J. (1993), 'Selling Issues to Top Management', *Academy of Management Review*, 18 (3): 397-428.

Dutton, J. E., Ashford, S. J., Lawrence, K. A., and Miner-Rubino, K.

(2002), 'Red light, green light: Making sense of the organizational context for issue selling', *Organization Science*, 13 (4): 355-369.

Dutton, J. E., Ashford, S. J., O'Neill, R. M., and Lawrence, K. A. (2001), 'Moves that matter: Issue selling and organizational change', *Academy of Management Journal*, 44 (4): 716-736.

Dutton, J. E., Ashford, S. J., Oneill, R. M., Hayes, E., and Wierba, E. E. (1997), 'Reading the wind: How middle managers assess the context for selling issues to top managers', *Strategic Management Journal*, 18 (5): 407-423.

Dwyer, G., Hardy, C., and Maguire, S. (2021), 'Post-Inquiry Sensemaking: The Case of the 'Black Saturday' Bushfires', *Organization Studies*, 42 (4): 637-661.

Edmondson, A. (1999), 'A safe harbor: Social psychological conditions enabling boundary spanning in work teams', *Research on Managing Groups and Teams, Vol 2 - 1999: Groups in Context*, 2 179-199.

Edmondson, A., Bohmer, R., and Pisano, G. (2001a), 'Speeding up team learning', *Harvard Business Review*, 79 (9): 125.

Edmondson, A. C. (1996), 'Learning from Mistakes is Easier Said Than Done: Group and Organizational Influences on the Detection and Correction of Human Error', *The Journal of Applied Behavioral Science*, 32 (1): 5-28.

Edmondson, A. C. (2003a), 'Framing for learning: Lessons in successful technology implementation', *California Management Review*, 45 (2): 34.

Edmondson, A. C. (2003b), 'Speaking up in the operating room: How team leaders promote learning in interdisciplinary action teams', *Journal of Management Studies*, 40 (6): 1419-1452.

Edmondson, A. C. (2004a), 'Learning from failure in health care: frequent opportunities, pervasive barriers', *Quality & Safety in Health Care*, 13 3-9.

Edmondson, A. C. (2004b), 'Learning from mistakes is easier said than done: Group and organizational influences on the detection and correction of human error', *The Journal of Applied Behavioral Science*, 40 (1): 66-90.

Edmondson, A. C. (2008), 'The competitive imperative of learning', *Harvard Business Review*, 86 (7-8): 60.

Edmondson, A. C. (2011), 'Strategies for Learning from Failure', *Harvard Business Review*, 89 (4): 48.

Edmondson, A. C. (2012), 'Teamwork on the Fly', *Harvard Business Review*, 90 (4): 72-80.

Edmondson, A. C. (2016), 'Wicked Problem Solvers', *Harvard Business Review*, 94 (6): 52-59.

Edmondson, A. C., Bohmer, R. M., and Pisano, G. P. (2001b), 'Disrupted routines: Team learning and new technology implementation in hospitals', *Administrative Science Quarterly*, 46 (4): 685-716.

Edmondson, A. C., Dillon, J. R., and Roloff, K. S. (2007), 'Three Perspectives on Team Learning Outcome Improvement, Task Mastery, and Group Process', *Academy of Management Annals*, 1 269-314.

Edmondson, A. C., and Harvey, J. F. (2018), 'Cross-boundary teaming for innovation: Integrating research on teams and knowledge in organizations', *Human Resource Management Review*, 28 (4): 347-360.

Edmondson, A. C., and Lei, Z. K. (2014), 'Psychological Safety: The History, Renaissance, and Future of an Interpersonal Construct', *Annual Review of Organizational Psychology and Organizational Behavior, Vol 1*, 1 23-43.

Edmondson, A. C., Winslow, A. B., Bohmer, R. M. J., and Pisano, G. P. (2003), 'Learning how and learning what: Effects of tacit and codified knowledge on performance improvement following technology adoption', *Decision Sciences*, 34 (2): 197-223.

Fayol, H. (1949, first pub. 1930), *Industrial and general administration* (London: Sir I. Pitman & Sons, Ltd.).

Feldman, M. S., and Pentland, B. T. (2003), 'Reconceptualizing organizational routines as a source of flexibility and change', *Administrative Science Quarterly*, 48 (1): 94-118.

Gaba, D. M. (2000), 'Structural and organizational issues in patient safety: A comparison of health care to other high-hazard industries', *California Management Review*, 43 (1): 83.

Galbin, A. (2021), 'Sensemaking in the Social Construction of Organization. A Powerful Resource in a Pandemic Context', *Postmodern Openings*, 12 (1): 308-318.

Gandhi, T. K., Kaplan, G. S., Leape, L., Berwick, D. M., Edgman-Levitan, S., Edmondson, A., Meyer, G. S., Michaels, D., Morath, J. M., Vincent, C., and Wachter, R. (2018), 'Transforming concepts in patient safety: a progress report', *BMJ Quality & Safety*, 27 (12): 1019-1026.

Garvin, D. A. (1984), 'What Does Product Quality Really Mean?', *Sloan Management Review*, 26 (1): 25-43.

Garvin, D. A. (1986), 'Quality Problems, Policies, and Attitudes in the United States and Japan – an Exploratory Study', *Academy of Management Journal*, 29 (4): 653-673.

Garvin, D. A. (1987), 'Competing on the 8 Dimensions of Quality', *Harvard Business Review*, 65 (6): 101-109.

Garvin, D. A. (1993), 'Building a Learning Organization', *Harvard Business Review*, 71 (4): 78-91.

Garvin, D. A. (1998), 'The processes of organization and management', *Sloan Management Review*, 39 (4): 33.

Garvin, D. A., Edmondson, A. C., and Gino, F. (2008), 'Is yours a learning organization?', *Harvard Business Review*, 86 (3): 109.

Gawande, A. (2009), *The checklist manifesto: how to get things right* (Henry Holt and Company, New York): 13.

Gawande, A. (2012), 'Two Hundred Years of Surgery', *New England Journal of Medicine*, 366 (18): 1716-1723.

Gawande, A. A., Thomas, E. J., Zinner, M. J., and Brennan, T. A. (1999), 'The incidence and nature of surgical adverse events in Colorado and Utah in 1992', *Surgery*, 126 (1): 66-75.

Gehman, J., Glaser, V. L., Eisenhardt, K. M., Gioia, D., Langley, A., and Corley, K. G. (2018), 'Finding Theory-Method Fit: A Comparison of Three Qualitative Approaches to Theory Building', *Journal of Management Inquiry*, 27 (3): 284-300.

Ghosh, M., and Sobek, D. K. (2015), 'A problem-solving routine for improving hospital operations', *Journal of Health Organization and Management*, 29 (2): 252-270.

Gilovich, T., Griffin, D., and Kahneman, D. (2002), *Heuristics and biases: The psychology of intuitive judgment*. Cambridge University Press.

Ginossar, Z., and Trope, Y. (1987), 'Problem-Solving in Judgment Under Uncertainty', *Journal of Personality and Social Psychology*, 52 (3): 464-474.

Gioia, D. A., and Chittipeddi, K. (1991), 'Sensemaking and Sensegiving in Strategic Change Initiation', *Strategic Management Journal*, 12 (6): 433-448.

Gioia, D. A., Corley, K. G., and Hamilton, A. L. (2013). 'Seeking Qualitative Rigor in Inductive Research: Notes on the Gioia Methodology', *Organizational Research Methods*, 16 (1): 15-31.

Gioia, D. A., and Mehra, A. (1996), 'Sensemaking in organizations - Weick, KE', *Academy of Management Review*, 21 (4): 1226-1230.

Glynn, M. A., and Watkiss, L. (2020), 'Of Organizing and Sensemaking: From Action to Meaning and Back Again in a Half-Century of Weick's Theorizing', *Journal of Management Studies*, 57 (7): 1331-1354.

Goh, T. N. (2010), 'Six Triumphs and Six Tragedies of Six Sigma', *Quality Engineering*, 22 (4): 299-305.

Goodman, P. S., Ramanujam, R., Carroll, J. S., Edmondson, A. C., Hofmann, D. A., and Sutcliffe, K. M. (2011), 'Organizational errors:

Directions for future research', *Research in Organizational Behavior: An Annual Series of Analytical Essays and Critical Reviews, Vol 31*, 31 151-176.

Grant, E. L. (1991), 'Statistical Quality Control in the World War II Years', *Quality Progress*, 24 (12): 31-36.

Groop, J., Ketokivi, M., Gupta, M., and Holmstrom, J. (2017), 'Improving home care: Knowledge creation through engagement and design', *Journal of Operations Management*, 53-56 9-22.

Gubrium, J., Holstein, J., Marvasti, A., and McKinney, K. D. (2012), *The SAGE Handbook of Interview Research: The Complexity of the Craft*, second edition.

Guest, G., Bunce, A., and Johnson, L. (2006), 'How many interviews are enough? An experiment with data saturation and variability'. *Field Methods*, 18 (1): 59-82.

Hales, B. M., and Pronovost, P. J. (2006), 'The checklist - a tool for error management and performance improvement', *Journal of Critical Care*, 21 (3): 231-235.

Hashem, A., Chi, M. T. H., and Friedman, C. P. (2003), 'Medical errors as a result of specialization', *Journal of Biomedical Informatics*, 36 (1-2): 61-69.

Haynes, A. B., Weiser, T. G., Berry, W. R., Lipsitz, S. R., Breizat, A. H. S., Dellinger, E. P., Herbosa, T., Joseph, S., Kibatala, P. L., Lapitan, M. C. M., Merry, A. F., Moorthy, K., Reznick, R. K., Taylor, B., Gawande, A. A., and Safe Surgery Saves Lives Study, G. (2009), 'A Surgical Safety Checklist to Reduce Morbidity and Mortality in a Global Population'. *New England Journal of Medicine*, 360 (5): 491-499.

Helmreich, R. L. (2000), 'On error management: lessons from aviation', *British Medical Journal*, 320 (7237): 781-785.

Herrmann, C. C., and Magee, J. F. (1953), 'Operations Research for Management', *Harvard Business Review*, 31 (4): 100-112.

Hodgkinson, G., and Starbuck, W. H. (2009), *The Oxford Handbook of*

Organizational Decision Making (Oxford).

Hoerl, R., Jensen, W., and de Mast, J. (2021), 'Understanding and addressing complexity in problem solving', *Quality Engineering*, 33 (4): 612-626.

Hollerer, M. A., Jancsary, D., and Grafstrom, M. (2018), 'A Picture is Worth a Thousand Words: Multimodal Sensemaking of the Global Financial Crisis', *Organization Studies*, 39 (5-6): 617-644.

Holmstrom, J., Tuunanen, T., and Kauremaa, J. (2014) *Logic for Design Science Research Theory Accumulation*.

Johnson, M., Burgess, N., and Sethi, S. (2020) 'Temporal pacing of outcomes for improving patient flow: Design science research in a National Health Service hospital', *Journal of Operations Management*, 66 (1-2): 35-53.

Juran, J. M. (1995), 'A History of Managing for Quality', *Quality Progress*, 28 (8): 125-129.

Just, K. S., Hubrich, S., Schmidtke, D., Scheifes, A., Gerbershagen, M. U., Wappler, F., and Grensemann, J. (2015), 'The effectiveness of an intensive care quick reference checklist manual - A randomized simulation-based trial', *Journal of Critical Care*, 30 (2): 255-260.

Kahneman, D. (2003), 'Maps of bounded rationality: Psychology for behavioral economics', *American Economic Review*, 93 (5): 1449-1475.

Kahneman, D. (2011), *Thinking, fast and slow*. New York: Farrar, Straus and Giroux.

Kahneman, D., Fredrickson, B. L., Schreiber, C. A., and Redelmeier, D. A. (1993), When More Pain is Preferred to Less – Adding a Better End', *Psychological Science*, 4 (6): 401-405.

Kahneman, D. and Tversky, A. (1979), 'Prospect Theory – Analysis of Decision under Risk', *Econometrica*, 47 (2): 263-291.

Kalkman, J. P. (2019), 'Sensemaking questions in crisis response teams', *Disaster Prevention and Management*, 28 (5): 649-660.

Kasatpibal, N., Senaratana, W., Chitreecheur, J., Chotirosniramit, N., Pakvipas, P., and Junthasopeepun, P. (2012) 'Implementation of the World Health Organization Surgical Safety Checklist at a University Hospital in Thailand', *Surgical Infections*, 13 (1): 50-56.

Kerrissey, M. J., Mayo, A. T., and Edmondson, A. C. (2021a), 'Joint Problem-Solving orientation in Fluid Cross-Boundary Teams', *Academy of Management Discoveries*, 7 (3): 381-405.

Kerrissey, M. J., Mayo, A. T., and Edmondson, A. C. (2021b) 'Joint Problem-Solving Orientation in Fluid Cross-Boundary Teams', *Academy of Management Discoveries*, 7 (3): 381-405.

Ketokivi, M., and Mantere, S. (2010), 'Two Startegies for Inductive Reasoning in Organizational Research', *Academy of Management Review*, 35 (2): 315-333.

Kish-Gephart, J. J., Detert, J. R., Trevino, L. K., and Edmondson, A. C. (2009), 'Silenced by fear: The nature, sources, and consequences of fear at work', *Research in Organizational Behavior, Vol 29: an Annual Series of Analytical Essays and Critical Reviews*, 29 163-193.

Klein, G. (2008), 'Naturalistic decision making', *Human Factors*, 50 (3): 456-460.

Klotz, F., and Edmondson, A. (2018), 'The Leadership Demands of "Extreme Teaming"', *Mit Sloan Management Review*, 59 (4): 42.

Kovács, G., and Spens, K. M. (2005), 'Abductive reasoning in logistics research', *International Journal of Physical Distribution & Logistics Management*, 35 (2): 132-144.

Kuhn, T. S. (2012), *The Structure of Scientific Revolutions*, (University of Chicago Press).

Lampel, J., Shamsie, J., and Shapira, Z. (2009), 'Experiencing the Improbable: Rare Events and Organizational Learning', *Organization Science*, 20 (5): 835-845.

Leape, L. L., and Berwick, D. M. (2005), 'Five years after "To err is human" - What have we learned?' *Jama-Journal of the American Medical Association*, 293 (19): 2384-2390.

Lee, N., and Lings, I. (2008), *Doing business research: A guide to theory and practice*. Sage.

Lewin, K. (1946), 'Action Research and Minority Problems', *Journal of Social Issues*, 2 (4): 34-46.

Liu, C. W., Vlaev, I., Fang, C., Denrell, J., and Chater, N. (2017), 'Strategizing with Biases: Making Better Decisions using the Mindspace Approach', *California Management Review*, 59 (3): 135-161.

Locke, E. A. (1982), 'The Ideas of Frederick W. Taylor: An Evaluation', *The Academy of Management Review*, 7 (1): 14-24.

Locke, K., Feldman, M., and Golden-Biddle, K. (2022), 'Coding Practices and Iterativity: Beyond Templates for Analyzing Qualitative Data', *Organizational Research Methods*, 25 (2): 262-284.

MacCarthy, B. L., Lewis, M., Voss, C., and Narasimhan, R. (2013), 'The same old methodologies? Perspectives on OM research in the post-lean age', *International Journal of Operations & Production Management*, 33 (7): 934-956.

MacDuffie, J. P. (1997), 'The road to "root cause": Shop-floor problem-solving at three auto assembly plants', *Management Science*, 43 (4): 479-502.

Mainthia, R., Lockney, T., Zotov, A., France, D. J., Bennett, M., St Jacques, P. J., Furman, W., Randa, S., Feistritzer, N., Eavey, R., Leming-Lee, S., and Anders, S. (2012), 'Novel use of electronic whiteboard in the operating room increases surgical team compliance with pre-incision safety practices', *Surgery*, 151 (5): 660-666.

Maitlis, S., and Christianson, M. K. (2014), 'Sensemaking in Organizations: Taking Stock and Moving Forward', *Academy of Management Annals*, 8 (1): 57-125.

Maitlis, S., and Sonenshein, S. (2010), 'Sensemaking in Crisis and Change: Inspiration and Insights From Weick (1988)', *Journal of*

Management Studies, 47 (3): 551-580.

Makary, M. A., and Daniel, M. (2016), 'Medical error-the third leading cause of death in the U.S.', *British Medical Journal*, 353 5.

Mathieu, J. E., Heffner, T. S., Goodwin, G. F., Salas, E., and Cannon-Bowers, J. A. (2000), 'The influence of shared mental models on team process and performance', *Journal of Applied Psychology*, 85 (2): 273-283.

Meehl, P. E. (1954), Clinical versus statistical prediction. (Minneapolis: University of Minnesota Press).

Meng, X., Rosenthal, R., and Rubin, DB (1992), 'Comparing correlated correlation coefficient', *Psychological Bulletin*, 111 172-175.

Michie, S., van Stralen, M. M., and West, R. (2011), 'The behaviour change wheel: A new method for characterising and designing behaviour change interventions', *Implementation Science*, 6 11.

Moaveni, S., and Sharma, I (2015), *Engineering Fundamentals: an Introduction to Engineering*,. 5e. (Cengage Learning).

Molina, G., Jiang, W., Edmondson, L., Gibbons, L., Huang, L. C., Kiang, M. V., Haynes, A. B., Gawande, A. A., Berry, W. R., and Singer, S. J. (2016), 'Implementation of the Surgical Safety Checklist in South Carolina Hospitals Is Associated with Improvement in Perceived Perioperative Safety', *Journal of the American College of Surgeons*, 222 (5): 725.

Monteiro, S. M., and Norman, G. (2013), 'Diagnostic Reasoning: Where We've Been, Where We're Going', *Teaching and Learning in Medicine*, 25 S26-S32.

Muller, S. D., Mathiassen, L., Saunders, C. S., and Kraemmergaard, P. (2017), 'Political Maneuvering During Business Process Transformation: A Pluralist Approach', *Journal of the Association for Information Systems*, 18 (3): 173-205.

Nawaz, H., Edmondson, A. C., Tzeng, T. H., Saleh, J. K., Bozic, K. J., and Saleh, K. J. (2014), 'Teaming: An Approach to the Growing

Complexities in Health Care', *Journal of Bone and Joint Surgery-American Volume*, 96A (21).

Nembhard, I. M., and Edmondson, A. C. (2006), 'Making it safe: The effects of leader inclusiveness and professional status on psychological safety and improvement efforts in health care teams', *Journal of Organizational Behavior*, 27 (7): 941-966.

Nowling, W. D., and Seeger, M. W. (2020), 'Sensemaking and crisis revisited: the failure of sensemaking during the Flint water crisis', *Journal of Applied Communication Research*, 48 (2): 270-289.

Onwuegbuzie, A., and Collins, K. M. T. (2007), 'A Typology of Mixed Methods Sampling Designs in Social Science Research', *Qualitative Report*, 12 281-316.

Patel, J., Ahmed, K., Guru, K. A., Khan, F., Marsh, H., Khan, M. S., and Dasgupta, P. (2014), 'An overview of the use and implementation of checklists in surgical specialities - A systematic review', *International Journal of Surgery*, 12 (12): 1317-1323.

Peerally, M. F., Carr, S., Waring, J., and Dixon-Woods, M. (2017), 'The problem with root cause analysis', *BMJ Quality & Safety*, 26 (5): 417-422.

Perrow, C. (2004), 'A personal note on Normal Accidents', *Organization & Environment*, 17 (1): 9-14.

Perrow, C. (2011), 'Fukushima and the inevitability of accidents', *Bulletin of the Atomic Scientists*, 67 (6): 44-52.

Perrow, C. (2013), 'Nuclear denial: From Hiroshima to Fukushima', *Bulletin of the Atomic Scientists*, 69 (5): 56-67.

Perrow, C. B. (2008), 'Complexity, catastrophe, and modularity', *Sociological Inquiry*, 78 (2): 162-173.

Petticrew, M., and Roberts, H. (2008), *Systematic Reviews in the Social Sciences: A Practical Guide*. John Wiley & Sons.

Pilbeam, C., Alvarez, G., and Wilson, H. (2012), 'The governance of supply networks: a systematic literature review', *Supply Chain Management-an International Journal*, 17 (4): 358-376.

Pisano, G. P., Bohmer, R. M. J., and Edmondson, A. C. (2001), 'Organizational differences in rates of learning: Evidence from the adoption of minimally invasive cardiac surgery', *Management Science*, 47 (6): 752-768.

Popper, K. (2005), *The Logic of Scientific Discovery*. Routledge.

Pratt, M. G. (2009), 'For the Lack of a Boilerplate: Tips on Writing Up (and Reviewing) Qualitative Research', *Academy of Management Journal*, 52 (5): 856-862.

Pratt, M. G., Kaplan, S., and Whittington, R. (2020), 'Editorial Essay: The Tumult over Transparency: Decoupling Transparency from Replication in Establishing Trustworthy Qualitative Research*', *Administrative Science Quarterly*, 65 (1): 1-19.

Pratt, M. G., Sonenshein, S., and Feldman, M. S. (2022), 'Moving Beyond Templates: A Bricolage Approach to Conducting Trustworthy Qualitative Research', *Organizational Research Methods*, 25 (2): 211-238.

Pugel, A. E., Simianu, V. V., Fluma, D. R., and Dellinger, E. P. (2015), 'Use of the surgical safety checklist to improve communication and reduce complications', *Journal of Infection and Public Health*, 8 (3): 219-225.

Rapley, T. J. (2001), 'The art(fulness) of open-ended interviewing: some considerations on analysing interviews', *Qualitative Research*, 1 (3): 303-323.

Rashid, F., Edmondson, A. C., and Leonard, H. B. (2013), 'Leadership Lessons from the Chilean Mine Rescue', *Harvard Business Review*, 91 (7-8): 113-+.

Redelmeier, D. A., and Kahneman, D. (1996) 'Patients' memories of painful medical treatments: Real-time and retrospective evaluations of two minimally invasive procedures', *Pain*, 66 (1): 3-8.

Roberts, K. H., and Bea, R. (2001), 'Must accidents happen? Lessons from high-reliability organizations', *Academy of Management Executive*, 15 (3): 70-78.

Roberts, K. H., Stout, S. K., and Halpern, J. J. (1994), 'Decision Dynamics in 2 High-Reliability Military Organizations', *Management Science*, 40 (5): 614-624.

Russ, S., Rout, S., Caris, J., Mansell, J., Davies, R., Mayer, E., Moorthy, K., Darzi, A., Vincent, C., and Sevdalis, N. (2015), 'Measuring Variation in Use of the WHO Surgical Safety Checklist in the Operating Room: A Multicenter Prospective Cross-Sectional Study', *Journal of the American College of Surgeons*, 220 (1): 1-U33.

Sahay, S., and Dwyer, M. (2021), 'Emergent Organizing in Crisis: US Nurses' Sensemaking and Job Crafting During COVID-19', *Management Communication Quarterly*, 35 (4): 546-571.

Sandberg, J., and Tsoukas, H. (2015), 'Making sense of the sensemaking perspective: Its constituents, limitations, and opportunities for further development', *Journal of Organizational Behavior*, 36 S6-S32.

Schaafstal, A. M., Johnston, J. H., and Oser, R. L. (2001) 'Training teams for emergency management', *Computers in Human Behavior*, 17 (5-6): 615-626.

Schildt, H., Mantere, S., and Cornelissen, J. (2020), 'Power in Sensemaking Processes', *Organization Studies*, 41 (2): 241-265.

Schmenner, R. W., and Swink, M. L. (1998), 'On theory in operations management', *Journal of operations management*, 17 (1): 97-113.

Schmenner, R. W., Van Wassenhove, L., Ketokivi, M., Heyl, J., and Lusch, R. F. (2009), 'Too much theory, not enough understanding', *Journal of Operations Management*, 27 (5): 339-343.

Schoemaker, P. J. H., Heaton, S., and Teece, D. (2018), 'Innovation, Dynamic Capabilities, and Leadership', *California Management Review*, 61 (1): 15-42.

Scott, C., and Medaugh, M. (2017), Axial Coding. In:

Silbey, S. S. (2016), Why Do So Many Women Who Study Engineering Leave the Field?

Singer, S. J., Molina, G., Li, Z. H., Jiang, W., Nurudeen, S., Kite, J. G., Edmondson, L., Foster, R., Haynes, A. B., and Berry, W. R. (2016), 'Relationship Between Operating Room Teamwork, Contextual Factors, and Safety Checklist Performance', *Journal of the American College of Surgeons*, 223 (4): 568-U557.

Smith, G. F. (1988), 'Towards a Heuristic Theory of Problem Structuring', *Management Science*, 34 (12): 1489-1506.

Snook, S. (2000), *Friendly fire: TheAaccidentalSshootdown of U.S. Black Hawks over Northern Iraq*.

Snow, D. A., Worden, S. K., Rochford, E. B., and Benford, R. D. (1986), Frame Alignment Processes, Micromobilization, and Movement Participation', *American Sociological Review*, 51 (4): 464-481.

Sole, D., and Edmondson, A. (2002), 'Situated knowledge and learning in dispersed teams', *British Journal of Management*, 13 S17-S34.

Sousa, R., and Voss, C. A. (2002), 'Quality management re-visited: a reflective review and agenda for future research', *Journal of Operations Management*, 20 (1): 91-109.

Starbuck, W. H. (2009), 'Cognitive Reactions to Rare Events: Perceptions, Uncertainty, and Learning', *Organization Science*, 20 (5): 925-937.

Stephens, K. K., Jahn, J. L. S., Fox, S., Charoensap-Kelly, P., Mitra, R., Sutton, J., Waters, E. D., Xie, B., and Meisenbach, R. J. (2020), 'Collective Sensemaking Around COVID-19: Experiences, Concerns, and Agendas for our Rapidly Changing Organizational Lives', *Management Communication Quarterly*, 34 (3): 426-457.

Stevenson, W. J. (2020) *Operations Management*. McGraw Hill.

Stolper, E., Van de Wiel, M., Van Royen, P., Van Bokhoven, M., Van der Weijden, T., and Dinant, G. J. (2011), 'Gut Feelings as a Third Track in General Practitioners' Diagnostic Reasoning', *Journal of General Internal Medicine*, 26 (2): 197-203.

Strauss, A., and Corbin, J. (1998) *Basics of QualitativeRresearch: Techniques and procedures for developing grounded theory, 2nd ed.* Thousand Oaks, CA: Sage Publications, Inc.

Sundstrom, E., Demeuse, K. P., are Futrell, D. (1990), "Work Teams – Applications and Effectiveness', *American Psychologist*, 45 (2): 120-133.

Sunstein, C., and Thaler, R. (2008), *Nudge: Improving decisions about Health, Wealth, and Happiness.*

Sutcliffe, K. M., Lewton, E., and Rosenthal, M. M. (2004), 'Communication failures: An insidious contributor to medical mishaps', *Academic Medicine*, 79 (2): 186-194.

Swendiman, R. A., Edmondson, A. C., and Mahmoud, N. N. (2019), 'Burnout in Surgery Viewed Through the Lens of Psychological Safety', *Annals of Surgery*, 269 (2): 234-235.

Thaler, R. H. (2015), *Misbehaving: The story of behavioral economics.*

Thaler, R. H., Tversky, A., Kahneman, D., and Schwartz, A. (1997), 'The effect of myopia and loss aversion on risk taking: An experimental test', *Quarterly Journal of Economics*, 112 (2): 647-661.

Thomassen, O., Espeland, A., Softeland, E., Lossius, H. M., Heltne, J. K., and Brattebo, G. (2011) 'Implementation of checklists in health care; learning from high-reliability organisations', *Scandinavian Journal of Trauma Resuscitation & Emergency Medicine*, 19 7.

Thompson, G. A. (2014), 'Labeling in Interactional Practice: Applying Labeling Theory to Interactions and Interactional Analysis to Labeling', *Symbolic Interaction*, 37 (4): 458-482.

Tranfield, D., Denyer, D., and Smart, P. (2003), 'Towards a methodology for developing evidence-informed management knowledge by means of systematic review', *British Journal of Management*, 14 (3): 207-222.

Treadwell, J. R., Lucas, S., and Tsou, A. Y. (2014), 'Surgical checklists: a systematic review of impacts and implementation', *BMJ Quality & Safety*, 23 (4): 299-318.

Tucker, A. L., and Edmondson, A. C. (2003), 'Why hospitals don't learn from failures: Organizational and psychological dynamics that inhibit system change', *California Management Review*, 45 (2): 55.

Tucker, A. L., Edmondson, A. C., and Spear, S. (2002a), 'When problem solving prevents organizational learnin',. *Journal of Organizational Change Management*, 15 (2): 122-137.

Tucker, A. L., Edmondson, A. C., and Spear, S. (2002b), 'When problem solving prevents organizational learning', *Journal of Organizational Change Management*, 15 (2): 122-137.

Tucker, A. L., Nembhard, I. M., and Edmondson, A. C. (2007), 'Implementing new practices: An empirical study of organizational learning in hospital intensive care units', *Management Science*, 53 (6): 894-907.

Tversky, A., and Kahneman, D. (1974), 'Judgment under Uncertainty – Heuristics and Biases', *Science*, 185 (4157): 1124-1131.

Tversky, A., and Kahneman, D. (1992), 'Advances in Prospect-Theory – Cumulative Representation of Uncertainty', *Journal of Risk and Uncertainty*, 5 (4): 297-323.

Valentine, M. A., and Edmondson, A. C. (2015), 'Team Scaffolds: How Mesolevel Structures Enable Role-Based Coordination in Temporary Groups', *Organization Science*, 26 (2): 405-422.

Valentine, M. A., Nembhard, I. M., and Edmondson, A. C. (2015), 'Measuring Teamwork in Health Care Settings A Review of Survey Instruments', *Medical Care*, 53 (4): E16-E30.

van Aken, J., Chandrasekaran, A., and Halman, J. (2016), 'Conducting and publishing design science research: Inaugural essay of the design science department of the Journal of Operations Management', *Journal of Operations Management*, 47-48 1-8.

van Hulst, M., and Tsoukas, H. (2021), 'Understanding extended narrative sensemaking: How police officers accomplish story work', *Organization*, 24.

Vivekanantham, S., Ravindran, R. P., Shanmugarajah, K., Maruthappu, M., and Shalhoub, J. (2014a), 'Surgical safety checklists in developing countries', *International Journal of Surgery*, 12 (1): 2-6.

Vivekanantham, S., Ravindran, R. P., Shanmugarajah, K., Maruthappu, M., and Shalhoub, J. (2014b), 'Surgical safety checklists in developing countries', *International Journal of Surgery*, 12 (1): 2-6.

Vogus, T. J., Sutcliffe, K. M., and Weick, K. E. (2010), 'Doing No Harm: Enabling, Enacting, and Elaborating a Culture of Safety in Health Care', *Academy of Management Perspectives*, 24 (4): 60-77.

Voss, C., Tsikriktsis, N., and Frohlich, M. (2002), 'Case research in operations management', *International Journal of Operations & Production Management*, 22 (2): 195-219.

Voss, C. A. (2020), 'Toward and Actionable and Pragmatic View of Impact', *Academy of Management Discoveries*, 6 (4): 538-539.

Voxted, S. (2017), '100 years of Henri Fayol', *Management Revue*, 28 (2): 256-274.

Weick, K. E. (1984), 'Small Wins – Redefining the Scale of Social Problems', *American Psychologist*, 39 (1): 40-49.

Weick, K. E. (1988), 'Enacted Sensemaking in Crisis Situations', *Journal of Management Studies*, 25 (4): 305-317.

Weick, K. E. (1990), 'The Vulnerable System – An Analysis of the Tenerife Air Disaster', *Journal of Management*, 16 (3): 571-593.

Weick, K. E. (1993), 'The Collapse of Sensemaking in Organizations – the Mann Gulch Disaster', *Administrative Science Quarterly*, 38 (4): 628-652.

Weick, K. E. (1995), *Sensemaking in organizations*. Thousand Oaks:

Sage Publications.

Weick, K. E. (1998), 'Improvisation as a mindset for organizational analysis', *Organization Science*, 9 (5): 543-555.

Weick, K. E. (2001), 'Friendly fire: The accidental shootdown of US black hawks over northern Iraq', *Administrative Science Quarterly*, 46 (1): 147-151.

Weick, K. E. (2004), 'Normal accident theory as frame, link, and provocation', *Organization & Environment*, 17 (1): 27-31.

Weick, K. E. (2007), 'The generative properties of richness', *Academy of Management Journal*, 50 (1): 14-19.

Weick, K. E. (2009), *Making Sense of the Organization: The impermanent organization, Vol 2*. (New York: John Wiley & Sons Ltd.)

Weick, K. E. (2010), 'Reflections on Enacted Sensemaking in the Bhopal Disaster', *Journal of Management Studies*, 47 (3): 537-550.

Weick, K. E. (2015), 'Ambiguity as Grasp: The Reworking of Sense', *Journal of Contingencies and Crisis Management*, 23 (2): 117-123.

Weick, K. E. (2020), 'Sensemaking, Organizing, and Surpassing: A Handoff*', *Journal of Management Studies*, 57 (7): 1420-1431.

Weick, K. E. & Putnam, T. (2006), 'Organizing for mindfulness - Eastern wisdom and Western knowledge', *Journal of Management Inquiry*, 15 (3): 275-287.

Weick, K. E. & Sutcliffe, K. A. (2006), 'Mindfulness and the quality of organizational attention', *Organization Science*, 17 (4): 514-524.

Weick, K. E. & Sutcliffe, K. M. (2015), *Managing the Unexpected: Sustained performance in a complex world*. (New York: John Wiley & Sons).

Weick, K. E., Sutcliffe, K. M., and Obstfeld, D. (2005), 'Organizing and the process of sensemaking', *Organization Science*, 16 (4): 409-421.

Weiser, A. K. (2021), 'The Role of Substantive Actions in Sensemaking During Strategic Change', *Journal of Management Studies*, 58 (3): 815-848.

Weiser, T. G., Haynes, A. B., Lashoher, A., Dziekan, G., Boorman, D. J., Berry, W. R., and Gawande, A. A. (2010), 'Perspectives in quality: designing the WHO Surgical Safety Checklist', *International Journal for Quality in Health Care*, 22 (5): 365-370.

Williams, E. A., and Ishak, A. W. (2018), 'Discourses of an Organizational Tragedy: Emotion, Sensemaking, and Learning After the Yarnell Hill Fire', *Western Journal of Communication*, 82 (3): 296-314.

Williams, T. A., Gruber, D. A., Sutcliffe, K. M., Shepherd, D. A., and Zhao, E. Y. F. (2017), 'Organizational Response to Adversity: Fusing Crisis Management and Resilience Research Streams', *Academy of Management Annals*, 11 (2): 733-769.

Wolbers, J. (2021), 'Understanding distributed sensemaking in crisis management: The case of the Utrecht terrorist attack', *Journal of Contingencies and Crisis Management*, 11.

Zicko, J. M., Schroeder, R. A., Byers, W. S., Taylor, A. M., and Spence, D. L. (2017), 'Behavioral Emergency Response Team: Implementation Improves Patient Safety, Staff Safety, and Staff Collaboration', *Worldviews on Evidence-Based Nursing*, 14 (5): 377-384.

Ziewacz, J. E., Arriaga, A. F., Bader, A. M., Berry, W. R., Edmondson, L., Wong, J. M., Lipsitz, S. R., Hepner, D. L., Peyre, S., Nelson, S., Boorman, D. J., Smink, D. S., Ashley, S. W., and Gawande, A. A. (2011), 'Crisis Checklists for the Operating Room: Development and Pilot Testing', *Journal of the American College of Surgeons*, 213 (2): 212-219.

APPENDIX Research Methods

This appendix provides additional detail describing the research methods used to develop the S.H.I.T. problem framework and the T.R.U.S.T. approach to augment traditional problem-solving.

Design Science Research: the combination of practice and theory

A Design Science Research (DSR) approach was used in this book. Like Action Research methods,[180] DSR seeks 'improvement-oriented knowledge'[181] using an approach of problem analysis, abductive solution design, empirical testing, and application of iterative learning loops. DSR draws from a synthesis of academic literature to design innovative propositions to address practitioner problems.[182] Learning from extant literature is applied to drive solutions through synthesis of different schools of thought to present propositions that are tested. The practical result of DSR is that knowledge captured from literature is changed or updated through empirical testing. DSR, by its nature, is an evidence-based methodology. See Figure 31 for the DSR process flow from practical problem to practical solution.

---

[180] Collatto et al. 2018
[181] van Aken et al. 2016: 2
[182] Holmstrom et al. 2014

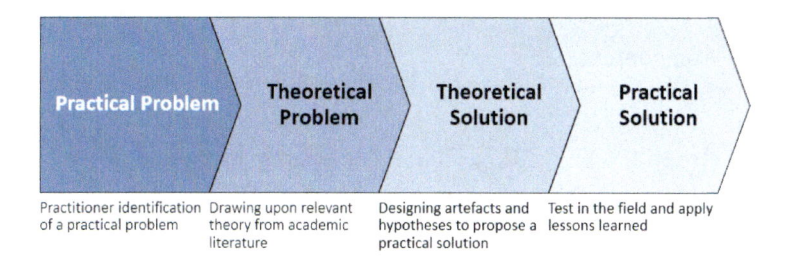

Figure 31 DSR process flow

DSR and CIMO-Logic for field research

Employing principles of DSR, this book assesses the current state-of-the-art knowledge through understanding of technical problem-solving literature and then devises solutions by incorporating the socio-psychological elements from Sensemaking and Psychological Safety to address the business challenges related to management of severe technical problems.

The research conducted for this book followed methods developed by Groop et al. (2017) in their adoption of Context, Intervention, Mechanism, and Outcomes[183] (CIMO) logic to guide the application of DSR. Users are instructed in the construction and performance of CIMO logic is follows:

> In this class of problematic Contexts, use this Intervention type to invoke these generative Mechanism(s), to deliver these Outcome(s)… In other words, the design proposition is comprised of a combination of interventions (I1…I11) that invoke particular generative mechanisms (M1 … M6) [from the

---

[183] Costa et al. 2018

literature review] to produce a particular outcome (O) in a specific context (C).[184]

Following Groop et al. (2017), the mechanisms identified from the literature review are used to design, create, and then test tools such as enhanced checklists and leadership routines to mitigate the undesirable effects and responses identified and categorised through interviews that undermine successful problem-solving during a severe event.

Interview approach and study participants

Semi-structured interviews

Interview candidates were selected to explore the dynamics around current technical problem-solving techniques and their failures to produce the optimal resolution in the context of severe events. The interviews provide the perspective of senior leaders responsible for engineering and operations in manufacturing and production industries, focusing on their lived experience of the problem-solving process and outcomes in respect of severe event problems.

Interviews are a qualitative research approach, employing procedure to support and explore research subjects' experience and behaviour through real-life cases to gain information and insight. '[G]ood cases help to identify the best stories, providing the basis for novel theory development.'[185] The research conducted for this book used interviews to explore the context of problem-solving under extreme conditions and to

---

[184] Denyer et al. 2008: 395-407
[185] MacCarthy et al. 2013: 939

identify empirical handling of severe events. Interviews 'can lead to new and creative insights, development of new theory, and have high validity with practitioners—the ultimate user of research'.[186] The use of interviews for this book created opportunity for detailed exploration of practical interventions employed by practitioners to mitigate the factors undermining successful problem-solving.

The purposive research population of this book[187] are senior and executive leaders, who work in the operations or engineering function at Fortune Global 500 companies.[188] Firms of this size universally adhere to standards established through ISO certification, which among other things, requires the use of specific problem-solving methods. Each interview examines one or more problems that occurred at these companies and the problem-solving efforts that followed.

Interview participants were all trained in technical problem-solving root cause analysis methods. In addition to formal training, each interviewee had a robust practical background in problem-solving. To recap, the three elements used to identify interview candidates were:

1. Senior and/or executive leaders at Fortune Global 500 firm
2. Employed in the operations or engineering functions
3. Involved personally in technical problems and problem-solving

---

[186] Voss et al. 2002: 195
[187] Bryman 2016
[188] https://fortune.com/global500/

Interview research method application

DSR principles were applied during the interviews to develop an understanding of the limitations of technical problem-solving in the context of severe event problems. Two stages of interviews were conducted:

1. Problem setting. Initial interviews explored the severe-event problem via the lived experience of senior and executive leaders to answer the questions: 'Is traditional problem-solving effective during severe events and if not, what factors limit its effectiveness?'

2. Solution implementation. Solution test interviews explored the effectiveness of the proposed solution design with key stakeholders of the pilot to answer the question: 'What were the outcomes when the traditional technical problem-solving approach was enhanced with the solution design and artifacts developed through the practical application of research conducted for this thesis project?'

Data gathering – Semi-structured interview protocol and process

Stage One data consists of 24 interviews with leaders working in the operations and/or engineering field. Interviewees were trained in formal technical problem-solving methods and each one had been professionally involved in one or more severe event arising from technical problems. Questions were designed to elicit insights from individuals with diverse functional positions, capabilities, and groups, because categorising and grouping through differences facilitates the comparative analyses and enables exploration of alternatives to refine the analysis based on what is working and not working with technical problem-solving. Interviews included Senior Executive, Vice-President, Director, and Senior Manager roles. The functional roles and position included working in a manufacturing plant, field service, supply chain, program management, quality, and health and safety.

Respondents were sent a pro forma interview guide 7-10 days in advance of interview, outlining the topics to be covered. All interviews were audio recorded and transcribed verbatim; each lasted between 60 and 120 minutes.

A semi-structured interview protocol was used to encourage respondents to provide rich descriptions[189] of their lived experience of problem-solving in the context of a severe event. Specifically, participants were asked to describe 'what they did', the types and kinds of problems they confronted in their role, the approaches or methods they utilised to address technical problems in their role, and whether there were differences across types of problems in practice that affect the outcome of the problem-solving endeavour.

Actual examples were examined in order to remain firmly in the phenomenon of interest (practical application) and not in the realm of ideas (theory). All interviews focused on at least one technical problem and questions prompted reflection to identify learning. For example, respondents were asked 'in hindsight, what could you have done differently to prepare for this problem?'

---

[189] Bearman 2019

The descriptions of real-world problem-solving efforts created a foundation for identifying which type of problems exceed the limits of traditional problem-solving, the elements of the traditional technical problem-solving process that need to be augmented, and possible methods for augmenting the process. An example of a question used from the interview topic guide is: 'Can you give me an example of when you have had a big problem at work?'

In addition, a prepared note sheet to guide the interview was used. Documentary data relating to the real-world problem-solving efforts was collected when available.

In summary, respondent narratives provided examples of severe events, failed problem-solving efforts, and opportunities to improve current problem-solving methods. The interviewees described in rich detail problem-solving in industry today, also revealing limitations and the types of problems where these limitations emerge, focusing on severe event case examples.

Limitations of the interview method

A possible drawback to using interviews as a research method is the potential for respondent bias arising from the relationship between the interviewer and the respondent. Of particular concern is the co-construction of narratives that can emerge due to factors such as power imbalance or the lack of understanding of context from interviewee to interviewer. To mitigate potential bias, Rapley (2001) prescribes 'awareness of the local context of the data production [as] central to analysing interviews'.[190] In this case, the author's prior professional experience as a problem-solving leader in the industrial sector helped to mitigate these risks.

The U.K. SAGE Handbook of Interview Research[191] cautions 'In-depth interviewing involves a certain style of social and interpersonal interaction …To be effective and useful, in-depth interviews develop and build on intimacy.'[192] The identity and background of the author was advantageous for the recruitment of highly experienced industry leaders, with whom a rapport was established through shared experiences of problem-solving.

Study participants and severe event identified

Technical problem-solving commonly occurs within the roles of operations and engineering professionals. The survey participants were selected to represent the perspectives of these different defined responsibilities. The 24 participants were selected using a purposive sampling approach. The interviewees represent senior leaders in a variety of manufacturing and production functions. Interviewees described a total of 27 unique technical problems. Figure 32 provides an overview of the 27 problem cases relied on for this research, including the job title of each person interviewed at the time of the interview, the type of company they work in, some context on the specific technical problem that was the focus of the interview, the traditional problem-solving process employed, and the problem resolution.

---

[190] Rapley 2001: 306
[191] Gubrium et al. 2012
[192] Gubrium et al. 2012: 2

Specific details on the problem background are removed from the book to preserve the strict confidentiality of the interview respondent.

	Case 1	Case 2	Case 3
Interviewee (title at time of interview)	1. Managing Director 2. Senior Quality Leader 3. Global Product Sales Director	1. Senior Project Management Leader 2. Senior HSE Leader 3. Senior Engineering Leader	1. Service Quality Leader 2. President and CEO 3. Regional Director of HSE (Health Safety Executive)
Problem-solving response	8D	8D	8D
Problem Resolution	8D used twice but halted. The problem was solved by technical experts using trial and error. The problem was documented in 8D format.	Stunted. 8D used twice – root cause never verified. The problem was documented in 8D format.	8D used multiple times.

	Case 4	Case 5	Case 6
Interviewee (title at time of interview)	Global Product Sales Director	Global Product Sales Director	Managing Director
Problem-solving response	8D	Trial and error	Six Sigma

Problem Resolution	8D tried and stopped. Problem ultimately solved with task force team of experts 'hands-on approach'	Problem successfully solved with team of experts using trial and error.	Six Sigma used at start of the project then stopped. Root Cause never fully agreed.

	Case 7	Case 8	Case 9
Interviewee (title at time of interview)	Executive Business Excellence Leader	Senior Quality Leader	Senior Quality Leader
Problem-solving response	8D	8D	8D
Problem Resolution	8D started then stopped before completion.	Stunted. 8D stopped and identified as not helpful. Problem solved by team of experts and then documented and presented to customer in 8D format.	Stunted. 8D stopped. Trial and error used to solve problem. Documented in 8D format post-problem for presentation. Consultants used in combination with internal experts to analyse and model failure.

	Case 10	**Case 11**	**Case 12**
Interviewee (title at time of interview)	Senior Manufacturing Manager	Quality and Operations Excellence Leader	Senior Vice-President, Operations
Problem-solving response	Trial and error	8D	Trial and error
Problem Resolution	No tool used. Preventive actions for the future never defined.	Trial and error used to solve problem. Documented in 8D format post-problem for presentation	Task force created. Due to sense of urgency, no formal problem-solving applied. Root cause of design issue never determined. Business cut the loss and recalled the product.

	Case 13	**Case 14**	**Case 15**
Interviewee (title at time of interview)	Senior Vice-President, Operations	Quality Director	Senior Executive Supply Chain
Problem-solving response	8D	8D	Emergency SWAT team
Problem Resolution	8D started then stopped. Trial and error used to solve	Six Sigma started then stopped.	Problem-resolved and plant production

			resumed. 8D report created to document event and learnings.
	problem. Documented in 8D format post-problem for presentation.		

	Case 16	Case 17	Case 18
Interviewee (title at time of interview)	Senior Materials and Planning Manager	President and CEO	Service Quality Leader
Problem-solving response	8D	8D	8D
Problem Resolution	8D started than abandoned. Problem solved using team of technical experts but documented for the customer in 8D format	Problem solved using team of technical experts but documented for the customer in 8D format	8D started than abandoned. Problem solved using team of technical experts but documented for the customer in 8D format

	Case 19	Case 20	Case 21
Interviewee (title at time of interview)	Service Quality Leader	Vice President of Sales	Vice President of Sales
Problem-solving response	8D	8D	8D
Problem Resolution	8D started than stopped.	8D started than stopped. Problem	8D started than stopped.

	Problem solved using team of technical experts but documented for the customer in 8D format	solved using team of technical experts but documented for the customer in 8D format	Problem solved using team of technical experts but documented for the customer in 8D format
	Case 22	**Case 23**	**Case 24**
Interviewee (title at time of interview)	President	Senior Health, Safety, and Environmental Leader	Senior Health, Safety, and Environmental Leader
Problem-solving response	Trial and error	Trial and error	8D
Problem Resolution	Resolution involved firings.	No formal problem-solving applied. No root cause analysis including lack of corrective and preventive actions for future improvements.	8D started and then stopped.
	Case 25	**Case 26**	**Case 27**
Interviewee (title at time of interview)	Senior HSE Leader	Head of Construction	Head of Construction
Problem-solving response	8D	8D	8D

| Problem Resolution | Stunted; 8D stopped. | 8D started than stopped. Problem solved using team of technical experts but documented for the customer in 8D format | 8D started than stopped. Problem solved using team of technical experts but documented for the customer in 8D format |

Figure 32 Interview overviews

All participants were employed by large global corporations that design, produce, and service manufactured goods. In order to minimise biases that might arise due to geographic or cultural differences, interviewees were selected and interviewed from around the globe including the U.S., U.K., France, Poland, Germany, Switzerland, Denmark, Mexico, Japan, People's Republic of China, and Taiwan. To ensure that participants delivered diverse and robust professional experience and represented professionals with substantial training and real-world practical experience with applied problem-solving, an array of senior and executive leaders was selected.

The number of interview participants was guided by the recommendation of Onwuegbuzie and Collins (2007: 289), 'sample sizes in qualitative research should not be so small as to make it difficult to achieve data saturation; at the same time, the sample should not be so large that it is difficult to undertake a deep analysis'. Interviews were continued until there was enough evidence that the author could successfully and consistently predict the conditions that would be present when additional interviewees described severe events where traditional problem-solving failed.

Data analysis

The 27 cases analysed are generated from 24 structured interviews using a protocol, with the examples being transcribed into a template for more analysis and comparison across the examples. Data was analysed following the 'case research in operations' method of turning qualitative data examples into cases to answer 'how and why' questions[193] prior to data coding and thematic analysis.[194] Analysis was iterative, involving refinements and revisions to the findings as they emerged over time.

All transcripts were coded in the qualitative software tool NVivo using a multi-step process. An initial open coding approach was followed by a more detailed axial coding method[195] concluded with '1st-order' and '2nd-order' evaluation aggregated into causal dimensions.[196]

The lessons from these interviews were employed to develop problem-solving artefacts that address the elements identified as contributing to the failure of traditional technical problem-solving tools in the context of a severe event. The unmet needs, experience, and lessons learned identified during this stage were employed to design solutions that augment existing approaches and compensate for the inadequacies of traditional approaches experienced throughout industry and across companies in the context of severe problems.

---

[193] Voss et al. 2002: 196
[194] Locke et al. 2022; Gioia et al. 2013
[195] Strauss and Corbin 1998; Scott and Medaugh 2017
[196] Gioia et al. 2013

Printed in Great Britain
by Amazon